Photomedicine and Stem Cells

The Janus face of photodynamic therapy (PDT) to kill cancer stem cells, and photobiomodulation (PBM) to stimulate normal stem cells

Photomedicine and Stem Cells

The Janus face of photodynamic therapy (PDT) to kill cancer stem cells, and photobiomodulation (PBM) to stimulate normal stem cells

Heidi Abrahamse

DST/NRF SARChI Chair: Laser Applications in Health, Laser Research Centre, Faculty of Health Sciences, University of Johannesburg, South Africa

Michael R Hamblin

Wellman Center for Photomedicine, Massachusetts General Hospital, Boston, MA USA Department of Dermatology, Harvard Medical School, Boston, MA USA Harvard-MIT Division of Health Sciences and Technology, Cambridge, MA, USA

Morgan & Claypool Publishers

Rights & Permissions
To obtain permission to re-use copyrighted material from Morgan & Claypool Publishers, please contact info@morganclaypool.com.

ISBN   978-1-6817-4321-9 (ebook)
ISBN   978-1-6817-4320-2 (print)
ISBN   978-1-6817-4323-3 (mobi)

DOI   10.1088/978-1-6817-4321-9

Version: 20171201

IOP Concise Physics
ISSN 2053-2571 (online)
ISSN 2054-7307 (print)

A Morgan & Claypool publication as part of IOP Concise Physics
Published by Morgan & Claypool Publishers, 1210 Fifth Avenue, Suite 250, San Rafael, CA, 94901, USA

IOP Publishing, Temple Circus, Temple Way, Bristol BS1 6HG, UK

I dedicate this work to all who left me with a treasure of everlasting memories of Greenside Road........
Heidi Abrahamse

Dedicated to the love of my life, my beautiful wife Angela.
Michael Hamblin

Contents

Preface

In this book we aimed to cover the field of photobiomodulation and photodynamic therapy applied both to normal stem cells and to cancer stem cells. The Janus face of photomedicine allows light to kill undesirable cells such as cancers and pathogenic infections, while protecting and stimulating injured or dysfunctional normal cells. The killing effect is obtained when visible light is combined with photosensitizers that produce reactive oxygen species. The stimulating effect takes advantage of the absorption of red and NIR light in the mitochondria in a process called photobiomodulation, leading to activation of transcription factors. Cancer stem cells are a subset of tumor cells thought to be related to tumor recurrence and resistance to cancer therapy. Photodynamic therapy as a treatment modality for cancer is highly effective yet highly underutilized despite its effective use as demonstrated in a variety of different types of cancer. Known to be quite resistant to conventional cancer treatment modalities, cancer stem cells can be treated by photodynamic therapy provided certain factors including fluence, photosensitizer concentration, uptake and cellular localization is controlled when applied. In particular, cancer stem cells possess a multi-drug efflux pump called ABCG2, that can lead to efflux of many (but not all) photosensitizers out of the cells. However, several inhibitors of ABCG2 are also known that have been shown to potentiate the PDT killing of cancer stem cells.

Photobiomodulation mainly affects cytochrome c oxidase located in the mitochondrial respiratory chain. However, stem cells are generally quiescent in their hypoxic niche, and have only low levels of mitochondrial activity. We propose a theory to explain why stem cells are highly responsive to light for example, when mitochondrial activity is stimulated by photobiomodulation, the cells' requirement for oxygen suddenly increases, and the cells are obliged to leave their hypoxic niche in search of more oxygen. When they emerge from their niche they are exposed to a barrage of biological cues that govern their proliferation and differentiation programs. Experimental evidence for this hypothesis has been provided by studies that show that irradiating the bone marrow can lead to repair of damage to the heart, brain and kidney. Furthermore, we explore the use of adipose derived stem cells as an easily and readily available source of stem cells for therapeutic use including disciplines such as dentistry and brain disorders.

This book is designed to show both the advantages of photobiomodulation for stem cell therapy and the possible use of photodynamic therapy to kill cancer stem cells. Readership that will benefit from this book includes scientists in the field of cell therapy, biochemistry and molecular biology while clinicians may benefit from understanding the basic principles underlying photobiomodulation as treatment modality.

Acknowledgement

HA would like to express her sincere gratitude to Anine Crous for her extensive assistance in sourcing the literature and compiling the final chapters.

Author biographies

Heidi Abrahamse

Heidi Abrahamse PhD Wits, (Biochemistry, Molecular Biology), is currently the Director of the Laser Research Centre, University of Johannesburg, and the Department of Science and Technology/ National Research Foundation SARChI Chair for Laser Applications in Health. Her research interests include photobiology and photochemistry with specific reference to photodynamic cancer therapy, stem cell differentiation, and wound healing. She has supervised 40 master's degrees, 15 doctorates, and 12 post-doctorate fellows, and has published over 100 peer-reviewed accredited journal publications, 42 accredited full paper proceedings, and 11 book chapters. She serves on the editorial boards of eight peer-reviewed internationally accredited journals while acting as reviewer for over 30 journals. She is also the Co-Editor in Chief of the international accredited journal *Photomedicine and Laser Surgery*.

Michael R Hamblin

Michael R Hamblin PhD is a Principal Investigator at the Wellman Center for Photomedicine, Massachusetts General Hospital, an Associate Professor of Dermatology, Harvard Medical School, and affiliated faculty at the Harvard-MIT Division of Health Science and Technology. He received his PhD in organic chemistry from Trent University in England. He directs a laboratory of around a dozen scientists who work in photodynamic therapy and photobiomodulation. He has published 376 peer-reviewed articles, is Editor or Associate Editor for ten journals and serves on NIH Study-Sections. He has an h-factor of 82 and >25 000 citations. He has authored/edited 11 proceedings volumes together with ten other major textbooks on photodynamic therapy and photomedicine. Dr Hamblin was honored by election as a Fellow of SPIE in 2011, received the first Endre Mester Lifetime Achievement Award in Photomedicine from NAALT in 2017 and the Outstanding Career Award from the Dose Response Society in 2018.

Photomedicine and Stem Cells

The Janus face of photodynamic therapy (PDT) to kill cancer stem cells, and photobiomodulation (PBM) to stimulate normal stem cells

Heidi Abrahamse and Michael R Hamblin

Chapter 1

Introduction

1.1 The Janus face of photomedicine

Janus (or Ianus) was the Roman god of beginnings, gates, transitions, time, doorways, passages, and endings. He is usually depicted as having two faces looking in opposite directions, since he looks to the future and also to the past (figure 1.1). It is thought that the month of January was named for Janus (Ianuarius) because he presided over the ending of one year and the beginning of the next. Janus also presided over the beginning and ending of conflict, and hence over war and peace. The doors of his temple were open in time of war, and closed to mark the onset of peace. The most famous temple to Janus in Rome is called the Ianus Geminus, or 'Twin Janus'. When its doors were open, neighboring cities knew that Rome was at war. As a god of transitions, he had functions pertaining to birth and to journeys and exchange.

Due to the two-faced nature of the Roman god, Janus has become inextricably linked with many aspects of biology and medicine. The concept of a process, mechanism, treatment, or drug having both good and bad effects is so widespread and pervasive that the analogy of Janus has been seized upon for a variety of reasons. Perhaps the best-known application of the Janus name in biology, is Janus kinase (JAK), which represents a family of intracellular, non-receptor tyrosine kinases that transduce cytokine-mediated signals via the JAK-STAT pathway. They were initially named 'just another kinase' 1 and 2 (since they were just two of a large number of discoveries in a PCR-based screen of kinases [1]), but were ultimately published as 'Janus kinase'. The name was taken from the two-faced Roman god of beginnings and endings because the JAKs possess two near-identical phosphate-transfer domains. One domain mediates the kinase activity, while the other negatively regulates the kinase activity of the first domain.

Some of the interesting and disparate medical topics that have been linked to Janus are demonstrated by the following publications: 'The Janus face of medicine'

Figure 1.1. Different depictions of the two-faced Roman god Janus.

discusses the likelihood of medicine to have both good and bad effects in general [2]; 'The Janus faces of botulinum neurotoxin: sensational medicine and deadly biological weapon' [3] is a good example of a therapeutic agent that can have dramatically opposite effects depending on dose; 'Depression and sickness behavior are Janus-faced responses to shared inflammatory pathways' discusses how the good, acute side of inflammation can lead to protection and resolution, while the bad, chronic side of inflammation can lead to clinical depression, a lifelong disorder [4]; in 'The Janus project: the remaking of nuclear medicine and radiology' [5] Larson discusses the conflict between the disciplines of nuclear medicine and radiology in the modern age of PET/CT.

We intend to show in this ebook that light therapy has many characteristics that can accurately be described as a 'Janus effect'. Depending on whether or not light is combined with a non-toxic photosensitizing dye (photodynamic therapy), and on the overall dose of light (energy density and power density), light is able to kill just about anything that is living, such as cancers, microorganisms, blood vessels, parasites, pests, unwanted tissues etc. On the opposite side of the coin (or perhaps on the other side of Janus' face), light of the correct wavelength and at the right dose (photobiomodulation) can have exactly the opposite effect, being able to heal, regenerate, protect, revitalize, and restore any kind of dead, damaged, stressed, dying, or degenerating cells, tissues, or organ systems. Between the bad destroying face of PDT, and the good healing face of PBM, it may well be concluded that 'all diseases can be treated with light'.

1.2 Photodynamic therapy—the killer

Photodynamic therapy (PDT) involves the combination of non-toxic dyes and harmless visible light in the presence of oxygen to produce highly toxic reactive oxygen species (ROS) that can kill unwanted cells such as cancer cells and pathogenic microorganisms [6]. PDT was discovered over one hundred years ago in 1900 when Oscar Raab (a medical student working with Herman von Tappeiner) examined the toxicity of acridine dyes to protozoa. Raab realized that the variable results he obtained in his experiments using acridine orange and paramecia could be attributed to variations in the level of ambient sunlight on the days when the experiments were performed. von Tappeiner and a dermatologist named Jesionek published the first report (just three years after Raab's initial paper) describing the use of 1% eosin solution painted on the skin with white light illumination to successfully treat patients with basal cell carcinoma of the skin. In the decade that followed, von Tappeiner's group showed that this 'photodynamic' effect required the simultaneous presence of a photosensitizer (PS), light, and oxygen to mediate cellular toxicity. Freiderich Meyer-Betz then performed the first human trial of hematoporphyrin-mediated PDT by injecting himself with 200 mg of the drug and standing in sunlight.

It was, however, not until 1978 that Thomas Dougherty and his colleagues at Roswell Park Cancer Institute (RPCI) reported the first large-scale trial of PDT in human patients with cancer [7]. In total, 113 primary and metastatic skin tumors were treated in 25 patients with diagnoses ranging from malignant melanoma to cutaneous T-cell lymphoma, to metastatic breast and endometrial cancers. Many of those tumors were recurrent or had progressed after multiple cytotoxic chemo-therapy agents and had failed local ionizing radiotherapy.

Since those early days PDT has continued to progress and has significantly widened the range of diseases and disorders that can be treated [6], in addition to cancer [8]. The ROS produced during PDT can rapidly and completely kill any type of pathogenic microorganism, whether it be different classes of bacteria, viruses, yeasts, filamentous fungi, or parasites, and can also remove unwanted tissue such as atherosclerotic plaque, undesirable blood vessels, and scars.

Figure 1.2 shows the mechanisms that operate during PDT. The ground-state PS has two electrons with opposite spins (singlet state) in the lowest-energy molecular orbital. Following the absorption of light (photons), one of these electrons is boosted

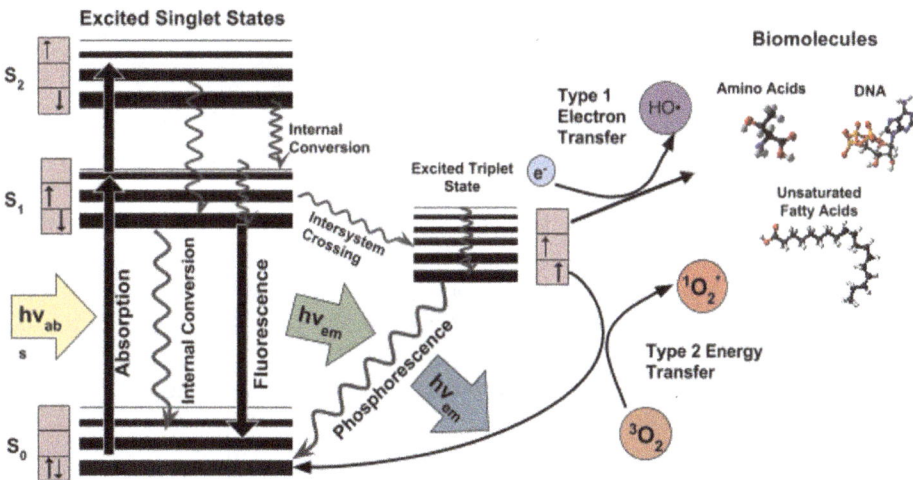

Figure 1.2. Jablonski diagram. Initial absorption of a photon by the ground state of the singlet PS gives rise to the short-lived first or second excited singlet states. The excited singlet state can lose energy by fluorescence, or by internal conversion to heat. Alternatively intersystem crossing can occur to give the long-lived triplet state. The PS triplet state is efficiently quenched by molecular oxygen (a triplet state) to give Type 2 (singlet oxygen) and Type I (superoxide and hydroxyl radical) reactive oxygen species. In the absence of oxygen the PS triplet state loses energy by phosphorescence. The singlet oxygen and hydroxyl radicals are potent oxidants that can damage biomolecules, amino acids, DNA and unsaturated fatty acids.

into a higher-energy orbital but retains its spin (first excited singlet state). This is a short-lived (nanoseconds) species and can lose its energy by emitting light (fluorescence) or by internal conversion into heat. The fact that most PSs are fluorescent has led to the development of sensitive assays to quantify the amount of PS in cells or tissues, and allows *in vivo* fluorescence imaging in living animals or patients to measure the pharmacokinetics and distribution of the PS. The excited singlet-state PS may also undergo the process known as intersystem crossing, whereby the spin of the excited electron inverts to form the relatively long-lived (microseconds to milliseconds) excited triplet-state that has both electron spins parallel. The long life of the triplet state is attributed to the relative rarity of other triplet molecules with which it can interact.

The PS excited triplet can undergo two broad kinds of reactions that are usually known as type I and type II (see the Jablonski diagram in figure 1.2). First, in a type I reaction, the triplet PS can gain an electron from a neighboring reducing agent. In cells these reducing agents are commonly either nicotinamide adenine dinucleotide (NADH) or NAD-phosphate, reduced form (NADPH). The PS is now a radical anion bearing an additional unpaired electron ($PS^{-\bullet}$). Alternatively, two triplet PS molecules can react together, involving intermolecular electron transfer, to produce a pair consisting of a radical cation and a radical anion. The PS radical anions may further react with oxygen to carry out an electron transfer to produce ROS, in particular superoxide anions. The damaging effects of superoxide are relatively mild, however, it can cause much more oxidative damage when it reacts with itself to produce hydrogen peroxide and oxygen, in the process called 'dismutation'. Hydrogen peroxide can add to an organic (carbon-containing) substrate and result in an oxidizing 'chain reaction'. This is common in the oxidative damage of fatty acids and other lipids. Hydrogen peroxide is important in the production of the highly reactive hydroxyl radical ($HO^{\bullet}$) by a second one-electron reduction mediated by $PS^{-\bullet}$. Superoxide will additionally react with nitric oxide ($NO^{\bullet}$) (also a radical) to produce peroxynitrite ($OONO-$), another highly reactive molecule that can oxidize many functional groups and can also nitrate proteins on tyrosine residues.

In a type II reaction, the triplet PS can transfer its energy directly to molecular oxygen (itself a triplet in the ground state), to form excited-state singlet oxygen. Type II processes are thought to best preserve the PS molecular structure in a repeatedly photoactivatable state, and in some circumstances a single PS molecule can generate 10 000 molecules of singlet oxygen. The PS can in some circumstances also react with the singlet oxygen it produces in a process known as oxygen-dependent photobleaching.

The singlet oxygen generated during type II photochemical reactions is believed to be the most important molecule responsible for PDT-induced cellular damage [9, 10]. However, because of the high reactivity and short half-life of singlet oxygen, only molecules and structures that are proximal to the area of its production (areas of PS localization) are directly affected by PDT. The half-life of singlet oxygen in biological systems is <40 ns, and, therefore, the radius of the action of singlet oxygen is of the order of 20 nm [11]. In singlet oxygen the spin restriction that reduces reaction rates of triplet oxygen with non-radicals is excluded. Consequently, singlet oxygen has an empty low-energy valence orbital that makes it a potent oxidizing/oxygenating agent compared to its ground-state counterpart. Singlet oxygen interacts with molecules via two essential mechanisms: physical interactions with quenchers or chemical reactions with biomolecules and scavengers. In the first phenomenon singlet oxygen transfers its excitation energy to an acceptor molecule (quencher) promoting it to an excited state while oxygen returns to ground state. The excited quencher can be cleaved, forming new products, or react with biomolecules and cause damage. Antioxidant molecules, such as carotenoids, can alternatively dissipate the energy quenched from singlet oxygen in the form of heat, preventing any sort of oxidative damage. Even though most carotenoids are very hydrophobic substances and exclusively accumulate in cell membranes, they are highly efficient in quenching singlet oxygen wherever it is produced [12].

There is an astonishing array of known compounds that can act as PSs in PDT. In the interests of brevity we can only give a few examples of their chemical structures here. Figure 1.3 shows the structure of Photofrin, the first PS to be clinically used for various cancers, and still one of the most common. Foscan or Temoporfin (m-tetrahydroxyphenylchlorin) is another PS that is commonly used to treat different types of cancers. Methylene blue is a phenothiazinium dye, which is the most commonly used PS to treat infections.

1.3 Cellular effects of PDT

1.3.1 Cellular targets of photo-oxidative damage

Singlet oxygen mainly reacts with molecules containing thiol groups or double bounds. Hence, it presents more restricted reaction patterns in relation to free radicals produced by type I reactions. Carbon and nitrogen atoms form double bonds sharing two electrons with opposite spin orientation. Note that ground-state oxygen cannot receive those two paired electrons because both low-energy orbitals are already occupied by one unpaired electron. Since singlet oxygen has an empty low-energy orbital, it has no restrictions to react with double bonds [12, 13].

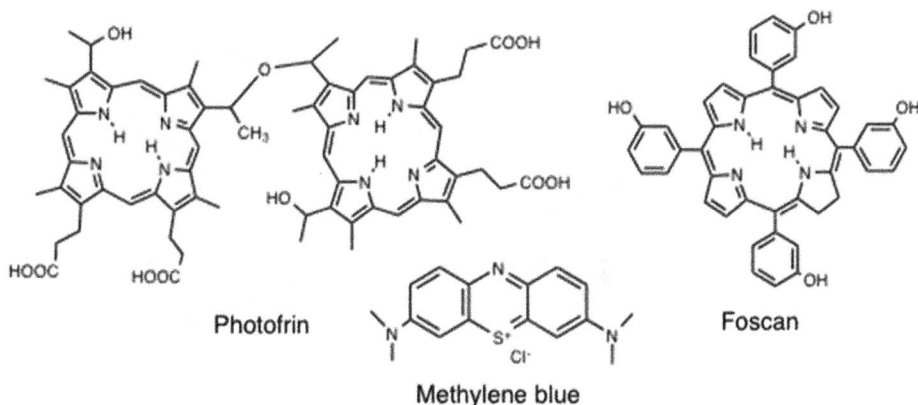

Figure 1.3. Chemical structures of three clinically used PSs: Photofrin, Foscan, and methylene blue.

Type 1 photodynamic reactions initiate via donation of one electron from the excited PS (^{3}PS*) to O_2 producing radicals ($R\bullet$), and peroxides that can yet produce more radicals. Radical species are typically paramagnetic because they carry one or more unpaired electrons in valence shell orbitals. This feature makes radicals particularly reactive. They are susceptible to react with several possible biomolecules passing the unpaired electron to another molecule also turning it into radical species or releasing another free radical. Therefore, a single radical formed can initiate chain reactions that lead to damage of several biomolecules. Radical reactions are rather unspecific and can attack lipids, sugars, proteins and nucleic acids forming radical moieties within these biomolecules. Such radical moieties then allow biomolecules to react with each other forming cross-links between them (e.g. lipid–lipid, lipid–protein, protein–DNA, etc). The radical addition to biomolecules through either hydroxylation and peroxidation or cross-link formation will potentially inhibit biomolecular function. As direct consequences, proteins are denatured, polysaccharides are cleaved, nucleic acids are damaged, and membranes become less fluid, more porous, and may even be disrupted. Proteins can also be targeted for proteasome degradation.

Lipid peroxidation represents a critical source of cellular damage that often leads to necrosis. Because hydroxyl radicals are extremely reactive non-polar species, they can rapidly abstract a hydrogen atom from saturated lipids forming lipid radicals ($L^{\bullet}$). Second, ground-state oxygen molecules can add to the lipid radical, forming peroxyl radicals that can subsequently react with other lipids that will propagate chain reactions until recombination or dismutation reactions take place. Since cell membrane consists of a double layer of phospholipids, closely interacting with each other and with membrane proteins, radical chain reactions can spread through the membrane causing serious damage that ultimately results in necrosis via membrane rupture [14]. To minimize the extent of this type of damage, nearly all cells in nature accumulate antioxidant molecules and enzymes in membranes, preventing the propagation of radical reactions [15, 16]. All amino acids, including when modified,

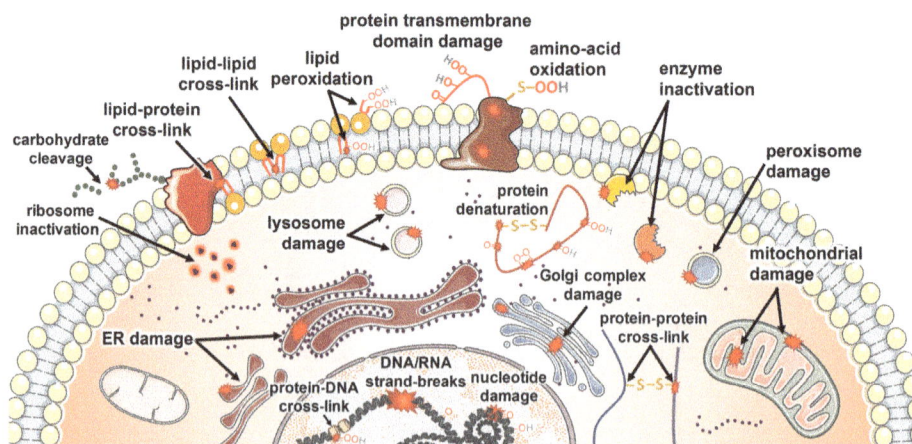

Figure 1.4. Cellular targets of PDT damage.

are susceptible to damage caused by radical reactions. Protein function is highly associated with its structural conformation, which is resultant of the chemical characteristics of each amino acid in the polymeric chain. Some amino acids are of course more important than others for protein structure and active sites; however, the oxidation of any amino acid in the chain can potentially alter protein conformation and consequently its function. In the case of membrane proteins, note that transmembrane domains are constructed with chains of hydrophobic amino acids. If these amino acids are oxidized they become more hydrophilic and may detach from the cell membrane to be solvated by water, either in the cytosol or the extracellular media [17].

1.3.2 Cell death pathways

Precisely why cells die when subjected to the PDT-generated ROS has been the subject of intense investigations in recent years. The discovery of programmed cell death or apoptosis has revolutionized the field of cytotoxic therapies in general and PDT in particular. The realization that apoptosis played a major role in embryonic development and in the immune system led to a search for ways in which this intrinsic property of cells could be harnessed to remove unwanted tissue and in particular cancerous cells. It should be emphasized that while ionizing radiation and chemotherapeutic drugs efficiently induce apoptosis in cancer cells, there are important differences between these modalities and PDT. Radiation and chemo-therapy tend to damage DNA and lead to apoptosis via cell-cycle checkpoints, growth arrest, and p53 activation. PDT, on the other hand, tends to operate via an acute stress response involving mitochondrial damage, cytochrome c release, and formation of an apotosome involving caspase.

A second discovery that has had a major impact on cell-based PDT research is the field of transcription factors. These are proteins that are often induced by acute insults and bind to certain regions on DNA, and therefore lead to transcription of genes and the consequent production of a multitude of proteins that affect cell function and cell death or survival.

The search to more closely define the molecular targets of PDT is inherently complex. Different cell types, different PSs, and different incubation and illumination conditions can all significantly alter the outcome of PDT. The question of whether there are particular cellular proteins that are more susceptible to oxidation by singlet oxygen or other PDT-generated ROS is just starting to be addressed. This question is closely related to the intracellular localization of the PS which governs its cellular targets (see figure 1.4). Another important question to be addressed is whether cancer cells are more susceptible to PDT-induced cell death than normal cells. The various mutations in cancer cells caused by the expression of oncogenes can lead to increased or decreased susceptibility to PDT. However, it is thought that many of the mutations that lead to resistance emerging in cancer cells exposed to radiation or chemotherapy do not lead to cross-resistance to PDT. If this proves to be the case then it would encourage the use of PDT against tumors that recur after conventional therapy, and repeated PDT treatments against the same tumor could also be envisaged. Another advantage of PDT may be that it does not lead to cumulative toxicity to the patient, and there is no known maximum cumulative dose as exists with both chemotherapy and radiation.

Apoptosis is a very complex, multi-step, multi-pathway cell death program that is inherited in every cell of the body [18]. It can be initiated either through the activation of a death receptor or the mitochondrial release of cytochrome c [19]. Both events eventually lead to activation of caspase cascades known as 'executioner caspases' such as caspase-3, -6, and -7 [20, 21]. The active executioner caspases cleave cellular substrates, which leads to the characteristic biochemical and morphological changes of dying cells [22]. Cleavage of nuclear lamins is followed by chromatin condensation and nuclear shrinkage; cleavage of the inhibitor of the caspase-activated deoxyribonuclease (DNase CAD) causes DNA fragmentation. Cleavage of cytoskeletal proteins leads to cell fragmentation and apoptotic body formation [23]. The apoptotic process is tightly controlled by various proteins [24, 25]. It is well known that the resistance of tumor cells to apoptosis might be an essential feature of cancer development; thus modulation of the key elements of apoptosis signaling may directly influence therapy-induced tumor cell death [26, 27].

1.4 Photobiomodulation—the healer

Light alone, when it is delivered at a relatively low irradiance and fluence, is called low-level light (laser) therapy (LLLT) and has almost the completely opposite effect to PDT. The modern terminology for LLLT has recently been changed to 'photo-biomodulation' (PBM) therapy, due to widespread uncertainties about just exactly what parameters constituted 'low level' [28]. LLLT (or PBM) was discovered in a rather serendipitous manner by Endre Mester in the 1960s. The discovery of the laser in 1960 by Theodore Maiman, had been recognized as a landmark in modern physics, and garnered an enormous amount of international attention [29]. In 1961 Leon Goldman started to experiment with the effect that these newly discovered laser beams had on the skin [30], and asked whether they could be used to perform 'bloodless surgery'. In the early 1960s Paul McGuff in Boston made medical history by using a laser beam to vaporize human cancer cells that had been transplanted into a hamster [31]. In 1965 Mester started laser research in the Semelweiss University Hospital in Budapest and tried to repeat McGuff's experiments by implanting tumor cells beneath the skin of laboratory rats and exposing them to the beam from a customized ruby laser. However the tumor cells were not destroyed by doses of what was presumed to be high-power laser energy, but instead, the skin incisions made to implant the cancer cells appeared to heal faster in the laser-treated animals, compared to the incisions of control animals that were not treated with light [32]. Moreover, the regrowth of hair on depilated rat skin was observed to be faster after exposure to his ruby laser [33]. After being initially puzzled by these contradictory findings, he realized that his custom-designed ruby laser was much weaker than he originally thought it to be, and instead of being photo-ablative against the tumor tissue, the low-power light stimulated the skin to heal faster and caused the hair to regrow.

The light is absorbed by chromophores in the mitochondria, such as cytochrome c oxidase, leading to production of more adenosine triphosphate (ATP) and activation of a host of cell signaling pathways leading to gene transcription [34]. Instead of killing cells LLLT stimulates cellular proliferation and prevents cells from dying [35]. The medical applications of LLLT range from serious diseases such as stroke, heart attack, spinal cord injury, and traumatic brain injury (TBI), to less serious indications such as wound healing, reduction of pain and inflammation, and depression [36].

1.4.1 PBM for wound healing and inflammation

The first demonstration of LLLT effectiveness was in non-healing wounds in patients in the early 1970s [37]. Since then many authors have published studies in both animal models and clinical trials showing positive effects. Our study in excisional wounds in mice showed that certain wavelengths (635 nm and 810 nm) were better than others and, moreover, there was a biphasic dose response with an optimum fluence of $2\,J\,cm^{-2}$ [38]. LLLT can be used with great benefit in numerous inflammatory disorders such as arthritis, carpal tunnel syndrome, and tendinitis. We showed that an 810 nm laser was very effective in a rat model of inflammatory arthritis caused by intra-articular injection of zymosan [39]. A sufficiently long illumination time was more important than fluence or irradiance in determining the benefit of the therapy.

1.4.2 PBM for traumatic brain injury and depression

Transcranial laser therapy has been studied for the treatment of stroke in animal models and in two clinical trials with impressive results [40]. A near-infrared laser (810 nm) can penetrate the scalp and skull and reach the cortical neurons in the brain. We have studied its possible benefits after TBI in mice. A single transcranial illumination of 4 h post-TBI produced significant improvements in neurological severity score that lasted for up to four weeks [41]. We also used 810 nm LED light delivered to the forehead as a possible treatment for severe anxiety and depression in a recent clinical trial [42]. Again, a single treatment had beneficial effects on Hamilton depression and anxiety scores that lasted for four weeks.

References

[1] Wilks A F 1989 Two putative protein-tyrosine kinases identified by application of the polymerase chain reaction *Proc. Natl Acad. Sci. USA* **86** 1603–7
[2] Boschung U 1996 The Janus face of medicine *Schweiz Med. Wochenschr.* **126** 507–11
[3] Osborne S L, Latham C F and Wen P J *et al* 2007 The Janus faces of botulinum neurotoxin: sensational medicine and deadly biological weapon *J. Neurosci. Res.* **85** 1149–58
[4] Maes M, Berk M and Goehler L *et al* 2012 Depression and sickness behavior are Janus-faced responses to shared inflammatory pathways *BMC Med.* **10** 66
[5] Larson S M 2011 The Janus project: the remaking of nuclear medicine and radiology *J. Nucl. Med.* **52** 3s–9s
[6] Hamblin M R and Mroz P 2008 *Advances in Photodynamic Therapy: Basic, Translational and Clinical* (Norwood, MA: Artech House)

[7] Cengel K A, Simone C B 2nd and Glatstein E 2016 PDT: what's past is prologue *Cancer Res.* **76** 2497–9

[8] Agostinis P, Berg K and Cengel K A *et al* 2011 Photodynamic therapy of cancer: an update *CA: Cancer J. Clin.* **61** 250–81

[9] Weishaupt K R, Gomer C J and Dougherty T J 1976 Identification of singlet oxygen as the cytotoxic agent in photoinactivation of a murine tumor *Cancer Res.* **36** 2326–9

[10] Henderson B W and Miller A C 1986 Effects of scavengers of reactive oxygen and radical species on cell survival following photodynamic treatment *in vitro*: comparison to ionizing radiation *Radiat. Res.* **108** 196–205

[11] Moan J and Berg K 1991 The photodegradation of porphyrins in cells can be used to estimate the lifetime of singlet oxygen *Photochem. Photobiol.* **53** 549–53

[12] Kearns D R 1971 Physical and chemical properties of singlet molecular oxygen *Chem. Rev.* **71** 32

[13] Bland J 1976 Biochemical effects of excited state molecular oxygen *J. Chem. Edu.* **53** 5

[14] Halliwell B and Chirico S 1993 Lipid peroxidation: its mechanism, measurement, and significance *Am. J. Clin. Nutr.* **57** 715s-724s 724s–725s

[15] Buettner G R 1993 The pecking order of free radicals and antioxidants: lipid peroxidation, alpha-tocopherol, and ascorbate *Arch. Biochem. Biophys.* **300** 535–43

[16] Halliwell B and Gutteridge J M C 2015 *Free Radicals in Biology and Medicine* 5th edn (Oxford: Oxford University Press)

[17] Davies M J 2016 Protein oxidation and peroxidation *Biochem. J.* **473** 805–25

[18] Pettigrew C A and Cotter T G 2009 Deregulation of cell death (apoptosis): implications for tumor development *Discov. Med.* **8** 61–3

[19] Rustin P 2002 Mitochondria, from cell death to proliferation *Nat. Genet.* **30** 352–3

[20] Perfettini J L and Kroemer G 2003 Caspase activation is not death *Nat. Immunol.* **4** 308–10

[21] Gougeon M L and Kroemer G 2003 Charming to death: caspase-dependent or -independent? *Cell Death Differ.* **10** 390–2

[22] Rathmell J C and Thompson C B 1999 The central effectors of cell death in the immune system *Annu. Rev. Immunol.* **17** 781–828

[23] Savill J and Fadok V 2000 Corpse clearance defines the meaning of cell death *Nature* **407** 784–8

[24] Okada H and Mak T W 2004 Pathways of apoptotic and non-apoptotic death in tumour cells *Nat. Rev. Cancer* **4** 592–603

[25] Cotter T G 2009 Apoptosis and cancer: the genesis of a research field *Nat. Rev. Cancer* **9** 501–7

[26] Igney F H and Krammer P H 2002 Immune escape of tumors: apoptosis resistance and tumor counterattack *J. Leukoc. Biol.* **71** 907–20

[27] Igney F H and Krammer P H 2002 Death and anti-death: tumour resistance to apoptosis *Nat. Rev. Cancer* **2** 277–88

[28] Anders J J, Lanzafame R J and Arany P R 2015 Low-level light/laser therapy versus photobiomodulation therapy *Photomed. Laser Surg.* **33** 183–4

[29] Maiman T H 1960 Stimulated optical radiation in ruby *Nature* **187** 493–4

[30] Goldman L, Blaney D J, Kindel D J Jr and Franke E K 1963 Effect of the laser beam on the skin *Preliminary report J. Invest. Dermatol.* **40** 121–2 1963

[31] Mcguff P E, Deterling R A Jr and Gottlieb L S 1965 Tumoricidal effect of laser energy on experimental and human malignant tumors *New Engl. J. Med.* **273** 490–2

[32] Mester E, Ludány G, Sellyei M, Szende B and Tota J 1968 The simulating effect of low power laser rays on biological systems *Laser Rev* **1** 3

[33] Mester E, Szende B and Gartner P 1968 The effect of laser beams on the growth of hair in mice *Radiobiol. Radiother. (Berlin)* **9** 621–6

[34] Chen A C-H, Arany P R and Huang Y-Y *et al* 2009 Low level laser therapy activates NF-kB via generation of reactive oxygen species in mouse embryonic fibroblasts *Mechanisms for Low-Light Therapy* vol 4 (San Jose)

[35] Huang Y Y, Chen A C, Carroll J D and Hamblin M R 2009 Biphasic dose response in low level light therapy *Dose Response* **7** 358–83

[36] Hashmi J T, Huang Y Y and Sharma S K *et al* 2010 Effect of pulsing in low-level light therapy *Lasers Surg. Med.* **42** 450–66

[37] Mester E, Szende B, Spiry T and Scher A 1972 Stimulation of wound healing by laser rays *Acta Chir. Acad. Sci. Hung* **13** 315–24

[38] Demidova-Rice T N, Salomatina E V, Yaroslavsky A N, Herman I M and Hamblin M R 2007 Low-level light stimulates excisional wound healing in mice *Lasers Surg. Med.* **39** 706-15

[39] Castano A P, Dai T and Yaroslavsky I *et al* 2007 Low-level laser therapy for zymosan-induced arthritis in rats: importance of illumination time *Lasers Surg. Med.* **39** 543–50

[40] Lampl Y 2007 Laser treatment for stroke *Expert Rev. Neurother.* **7** 961–5

[41] Wu Q, Huang Y-Y and Dhital S *et al* 2010 Low level laser therapy for traumatic brain injury *Proc. SPIE* 7552

[42] Schiffer F, Johnston A L and Ravichandran C *et al* 2009 Psychological benefits 2 and 4 weeks after a single treatment with near infrared light to the forehead: a pilot study of 10 patients with major depression and anxiety *Behav. Brain Func.* **5** 46

Photomedicine and Stem Cells
The Janus face of photodynamic therapy (PDT) to kill cancer stem cells, and photobiomodulation
(PBM) to stimulate normal stem cells
Heidi Abrahamse and Michael R Hamblin

Chapter 2

Cancer stem cells

Cancer stem cells (CSCs) have dominated the cancer and stem cell research sector in recent years. CSCs are a subset of tumor cells, found at the center of a tumor mass, possessing the ability to self-renew and differentiate, both characteristics thought to be related to tumor maintenance and carcinogenesis. In addition, they have the ability to de-differentiate which is thought to be a characteristic related to cancer metastases. CSCs have been identified as drivers of therapeutic resistance of cancers, as well as cancer recurrence.

2.1 Evidence for the existence of CSCs and their characteristic markers

Recent studies suggest that CSCs are immortal tumor-initiating cells that can self-renew and have pluripotent capacity [1]. CSCs have been identified in multiple malignancies, including leukemia and various solid cancers, and are thought to be the basis for tumor initiation, development, metastasis, and recurrence. Although Bruce *et al* observed that 1%–4% of lymphoma cells can form colonies *in vitro* or initiate carcinoma in mouse spleen [2], the first compelling evidence of CSCs was provided by Bonnet and Dick in 1997 [3] showing that CD34+CD38− cells from acute myeloid leukemia (AML) patients initiate hematopoietic malignancy in NOD/SCID mice. These cells possessed the ability to self-renew, proliferate and differentiate [3]. The first report of CSCs in solid cancer was provided in 2003 by Al-Hajj, demonstrating CSCs in breast cancer [4].

CSCs have now been demonstrated in solid tumors, including lung cancer, colon cancer, prostate cancer, ovarian cancer, brain cancer, and melanoma, among others [5], but the most compelling evidence of CSC identity comes from serial transplantation of cellular populations into animal models. This required the development of orthotropic transplantation assays. CSC-containing populations must be able to re-establish the phenotypic heterogeneity evident in the primary tumor and

doi:10.1088/978-1-6817-4321-9ch2   2-1

have the self-renewing capability of serial passaging. Significant technical difficulties are associated with the isolation of CSCs from epithelial and other solid tumors. During xenotransplantation, incomplete immunosuppression or species-specific differences in cytokines or growth factors cause problems. Implantation of tumor cells into a normal niche of syngeneic models does not imitate the tumor environment itself. Cell surface markers are useful for the isolation of subsets enriched for CSCs in solid tumors from patients. Markers such as CD133 (PRom1), CD44, CD24, epithelial cell adhesion molecule (epCAm; also known as epithelial specific antigen (eSA) and TACSTD1), THY1 and adenosine triphosphate (ATP)-binding cassette B5 (ABCB5), as well as Hoechst33342 exclusion by the side population (SP) cells have been identified [6]. Markers of CSCs in solid tumors from patients include the following: aBcG5, a member of the ABC family and transports sterol and other lipids; ABCG2, a breast cancer resistance protein, which is a multi-drug transporter; ABCG5, which confers doxorubicin resistance; aLDH1, the aldehyde dehydrogenase (ALDH) family of enzymes, which catalyzes the oxidation of aliphatic and aromatic aldehydes to carboxylic acids; cD24 (HSa), a highly glycosylated glycosylphosphatidylinositol-anchored adhesion molecule that co-stimulates B- and T-cells and is a ligand for P-selectin; cD44 (PGP1), an adhesion molecule with a pleiotropic role in signaling, migration, and homing; cD90 (THY1), a glycosylphosphatidylinositol-anchored membrane glycoprotein involved in signal transduction and stem cell differentiation; cD133 (prominin 1), a multi-domain glycoprotein organizing plasma membrane topology that is expressed on CD34+ stem and progenitor cells in fetal liver, endothelial precursors, fetal neural stem cells, and developing epithelium, and has been detected by its glycosylated epitope in the majority of studies; epCAm (eSA, TrOP1), a homophilic Ca2+-independent cell adhesion molecule expressed on the surfaces of most epithelial cells; and Hoechst SP, which has a SP phenotype due to the Hoechst33342 efflux pump present on the plasma membrane in diverse cell types.

Bonnet and Dick [3] isolated a subpopulation of leukemic cells that expressed a specific surface marker CD34, but not the CD38 marker. This provided evidence of tumorogenicity when transplanted into mice, as it phenotypically resembled the patient's original tumor. This method of identifying CSC markers has become the standard [7]. It is important to identify and target CSCs for successful cancer therapies in order to avoid a cancer relapse. Several studies have defined CSCs by using distinctive biomarkers in a variety of human cancers [8]. Certain markers may be more specific for one type of CSC than another, and some CSCs can be defined by the absence of certain markers [9]. CSCs have been isolated using specific markers for normal stem cells of the same organ. Characteristic features of CSC populations isolated from solid tumors are provided in table 2.1.

Common cell surface markers, such as CD133 and CD44, can fractionate CSCs in solid tumors. It is not known if they represent surrogate markers or whether they simply regulate CSC function. However, none of these markers are exclusively expressed by solid tumor CSCs [6].

There is substantial evidence for the existence of CSCs in carcinomas despite the fact that CSC fractions in solid tumors are impure populations. CSC frequencies

Table 2.1. Stem cell markers for common cancers.

Cancer	Biomarker	Ref.
Breast	CD44, CD49, CD133, epCAm, ALDH, CD24	[6, 9]
Brain	CD133	[6]
Colon	CD133, epCAm, CD44	[6]
Head and neck	CD44, CD133, ALDH1	[6, 9]
Pancreas	CD44, CD24, eSA, CD133	[6]
Lung	ALDH1, CD44, CD133, CD116, epCAm	[6, 9]
Liver	CD90$^+$	[6]
Melanoma	ABCB5$^+$	[6]
Mesenchymal	SP (Hoechst dye)	[6]
Acute myeloid leukemia	CD34	[9]
Ovarian cancer	CD133, CD44, CD117, CD24	[9]
Colorectal cancer	CD133, CD24, CD26, CD29, CD44, CD166, epCAm, ALDH	[9]
Prostate cancer	CD44, CD133, ALDH1	[10]
Cervical cancer	ALDH1, SOX2, Bcrp1, CD49, CD133	[11]

differ between tumors of the same subtype, but they occur at a higher frequency in more aggressive tumors. More definitive markers are required, since many antigens are widely expressed outside the putative CSC population. CD133 is a marker for many different types of CSCs, as well as normal stem cells, which may indicate their role in sustaining the stem cell phenotype. CD133 is expressed widely in the colon, however, whether CD133 promoter activity correlates with CD133 surface expression and whether the expression of glycosylated CD133 (AC133 epitope) is restricted to CSCs is still questioned. The lack of similarity between different markers reported for CSCs within a given tumor type, such as those in breast and pancreatic tumors, may be as a result of impure populations. Gene expression analyses, proteomics and generation of monoclonal antibodies against cell surface antigens may assist in identifying more markers. A combination of refined markers should improve purity [6].

2.2 Role in tumor recurrence

Solid tumors remain a cancer burden and a therapeutic challenge. The CSC hypothesis stems from the fact that stem cells are defined by their capacity for self-renewal and differentiation, and all tissues in the body are derived from organ-specific stem cells. CSC differentiation is restricted to cells of a particular organ. The hypothesis states that cancer progression is governed by a subpopulation of cancer-initiating cells that have stem-cell-like properties and the potential for unlimited replication and tumor initiation [12]. CSCs are also thought to be the reason for relapse following therapy [10]. The hypothesis states two related, but separate, issues: (i) the cellular origin of tumors and (ii) tumors are driven by cellular components that display 'stem cell properties'. Today, the validity of the hypothesis has been confirmed, and made possible, by advances in research methods such as fluorescence-activated cell sorting, coupled with advances in stem cell biology and the development of new xenotransplantation animal models [13]. These advances have aided in the identification of several types of CSCs from as far back as 1994 when CSC were identified in myeloid leukemia [14]. Since then CSCs have been isolated from various cancers including: breast cancer [4], prostate cancer [15], pancreatic cancer [16], head and neck cancer [17], lung cancer [18], hepatocellular carcinoma [19], and most recently renal CSCs [20].

The CSC hypothesis provides a cellular mechanism to account for the therapeutic deficiencies and dormant behavior exhibited by many tumors. Solid tumors are organized and sustained by a distinct subpopulation of CSCs. Evidence for the CSC hypothesis was supported in mouse models of epithelial tumorigenesis, although alternative models of heterogeneity also seem to apply. The relevance of CSCs remains a question, although early results indicate that specific targeting may be involved [6].

Clinically characterization through histology and expression of specific markers, as well as gene expression profiling, can define distinct tumor subtypes. The cellular origins of most solid tumors are unknown, but may be related to different subtypes indicative of distinct cells of origin when the tumor initiates. Genetic and epigenetic mutations as well as interactions between tumor cells and stroma, inflammatory cells, and recruited vasculature, contribute significantly to the tumorigenic process. Metastasis and tumor dormancy occur in many solid tumors but the mechanisms of these processes remain unknown [6].

It is postulated that most tumors arise from one single cell that develops into a heterogeneous population, although two distinct models have been proposed to account for tumor growth and heterogeneity within tumors. The CSC model includes initiation, progression, metastasis, and recurrence of cancer depending on rare stem cells. The heterogeneity and hierarchy between the cells of a tumor result from asymmetric division of CSCs, suggesting that tumors are highly hierarchical with a unique self-renewing population of cells at the top of the hierarchy. All other cells comprising the tumor bulk are derived from differentiated CSCs [21].

The clonal evolution model (CE model) [22] proposes that all tumor cells contribute to tumor maintenance with different capacities, namely the intercellular variation is attributed to sub-clonal differences that result from genetic and/or epigenetic changes during cancer development. The CSC model emphasizes functional heterogeneity without considering the existence of intracellular genetic variation or genetically diverse sub-clones, while the CE model highlights the genetic heterogeneity but ignores the functional variation within individual genetic sub-clones. Accumulating evidence suggests that neither the CSC model nor the CE model should be discarded, since research combining functional assays with genetic analysis to examine genetic diversity of tumor propagating cells or tumor-initiating cells in both leukemia and solid cancer have been performed [23]. Anderson and co-workers showed the genetic diversity of cancer propagating cells within individual ETV6-RUNX1-positive acute lymphoblastic leukemia (ALL) patients and that the genetic diversity and relative dominance of sub-clones vary with the development of disease [24]. Notta *et al* [25] showed that individual Bcr-Abl + ALL samples consist of genetically distinct sub-clones related by a complex evolutionary process. Genetically diverse sub-clones possess variable aggressive properties at the time of diagnosis, indicating that CSCs exist and evolve over time. An ancestral clone leads to two clonal lineages that evolve independently and each clone acquires diverse genetic aberrations. One clone emerges as the dominant diagnostic clone, while the other clone gives rise to the predominant clone containing additional mutations at relapse [25].

Functional heterogeneity of cells rendering different tumor subtypes within the tumor population itself as well as cells exhibiting distinct proliferative and differentiative capacities (referred to as tumor heterogeneity) are noted [26]. Cellular mechanisms underlying tumor heterogeneity are investigated intensely in the field of cancer biology. Southam and Brunschwig [27], provided evidence of heterogeneity within tumors by autologous transplantation of malignant cells from patients with different carcinomas into subcutaneous tissue. Studies of spontaneous mouse leukemias and lymphomas showed that the frequency of tumor-propagating cells ranged from 1% to the majority of cells [28, 29]. Functional heterogeneity among cancer cells derived from lung, ovary, and brain tumors was also evident in colony-forming assays *in vitro* [30].

2.3 Anti-CSC approaches

Current therapeutic interventions against cancer are limiting and frequently lead to treatment failure. Resistance to chemotherapy and radiotherapy seems to be the common cause of treatment failure in multiple malignancies. Strategies that are not selective against CSCs are harmful to healthy tissues, and patients face the risk of recurrence and metastasis because most therapies cannot eliminate CSCs [31]. CSC populations are more resistant to conventional cancer therapies than non-CSC populations. Elimination of CSCs is of vital importance when treating malignant diseases [32–34]. Many novel therapeutic modalities have been designed with the aim of eliminating CSCs and modifying the niches that support these cells. Different surface markers and changes in signaling pathways might be novel therapeutic targets as well as multiple potential CSC therapeutic targets, including the ABC superfamily, anti-apoptotic factors, detoxifying enzymes, DNA repair enzymes, and distinct oncogenic cascades (such as the Wnt/β-catenin, hedgehog, EGFR, and Notch pathways) have been suggested [34, 35]. Some therapeutic strategies successfully kill CSCs, while others are still under preclinical and clinical evaluation. Novel CSC-targeted therapeutic interventions are shown in figure 2.1 [5].

Four different strategies summarize the current popular ideas. By selectively targeting surface markers of CSCs (yellow area), more accurate results and fewer side effects can be achieved. Modern molecular biology techniques have identified more crucial signal elements and pathways (purple area). By interrupting significant pathways, specific characteristics of CSCs are suppressed. Third generation molecular drugs inhibiting the ABC cassette (orange area) are currently under clinical investigation. Finally, avoiding growth of blood vessels by altering the pH of the

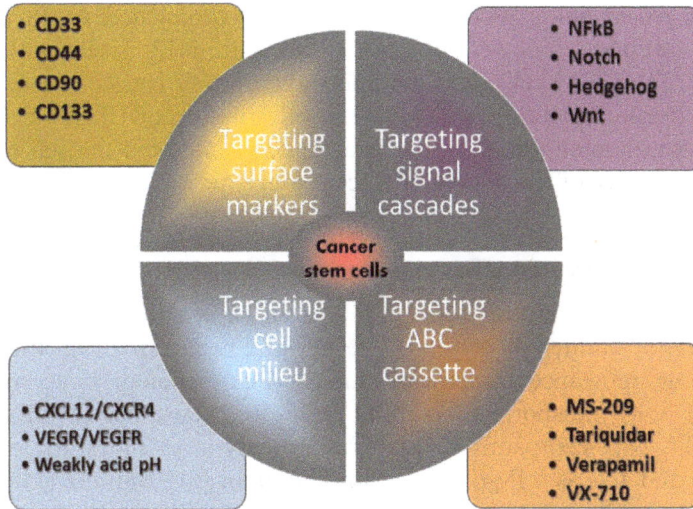

Figure 2.1. Therapies targeting CSCs (modified from [5]).

tumor microenvironment nourishing CSCs also shows promise as therapeutic intervention (blue area).

2.3.1 Targeting cellular surface markers

The development of monoclonal antibodies to target CSCs is a new strategy that has been suggested. Since CD33 is highly expressed in most AML cells, gemtuzumab ozogamicin, a humanized anti-CD33 mouse monoclonal antibody conjugated to the cytotoxic agent calicheamicin, has been developed and used to treat AML [36]. Markers differentially expressed between normal stem cells and CSCs, including CD44 [37], IL-3R [38], and the immunoglobulin mucin TIM-3 [39], have been utilized to specifically target leukemia stem cells in human AML.

Treatment with antibodies against cell surface molecules decreases leukemic genicity and eradicates CSCs in mice while antibodies against CD47, expressed at much higher levels in ALL than in normal cells, effectively kill leukemia stem cells [40]. In treating MCF-7 breast cancer, anti-CD44 antibody-conjugated gold nano-rods have been used to target and photo-ablate CD44+ cells. Targeted cells absorb near-infrared (NIR) light, which results in increased local temperature at the designated location [41].

An anti-CD90 antibody has also been reported to protect ATP-derived multi-potent stromal cells from differentiation into chondro-, osteogenic, or adipo-lineages [42]. CD133 is expressed in many types of CSCs, including lung cancer, breast cancer, and glioma, and yet patients with high levels of CD133 show poor clinical outcomes [43] as demonstrated by Mehra [44] where increased CD133 mRNA expression correlated with a high risk of death in colon, prostate and head and neck cancer. Therefore, although therapy against CD133 indicates a promising strategy for cancer treatment, significant research needs to be conducted to confirm its prospects. A recent study with carbon nanotubes conjugated to an anti-CD133 monoclonal antibody followed by irradiation with NIR laser light selectively targeted CD133+ GBM cells, and the photo-thermolysis caused by the nanotubes induced cell death in the targeted cells [45]. Interruption of CD133 expression by hairpin RNA in human glioblastoma neurospheres impaired the self-renewal and tumorigenic capacity of the neurosphere cells [46].

2.3.2 Targeting ATP-driven efflux transporters

Chemoresistance are often caused by ATP-driven pumps affecting anti-tumor drug efflux [31, 47, 48]. Numerous methods to evade and neutralize drug efflux pumps to overcome drug resistance have been studied. Pharmacological agents that can interact with ABC transporters to inhibit multi-drug resistance (MDR) have been developed [49, 50]. Verapamil is used in SP analysis to block the exclusion of Hoechst dye and acts as a P-gp efflux pump inhibitor [51]. Simultaneous treatment with Verapamil and anti-tumor drugs, such as Dox, Paclitaxel, or Vincristine, show promise. Khdair *et al* [51] delivered methylene blue and Dox simultaneously into BALB/c mice with syngeneic JC adenocarcinoma tumors, and discovered an increased accumulation of drugs within the lesion, enhanced tumor cell apoptosis, suppressed cancer cell proliferation, impaired tumor growth, and significantly improved animal survival.

A number of novel ABC transporter inhibitors that suppress both P-gp and MRP1, such as MS-209, VX-710, and tariquidar, have been identified [49, 52, 53]. Trials with MS-209 showed results in overcoming drug resistance in breast cancer [52] as well as solid cancers (ClinicalTrials.gov Identifier: NCT00004886). Tariquidar is an inhibitor that in combination with docetaxel is studied for the treatment of recurrent or metastatic ovarian, cervical, lung, and kidney malignancies, and the combination of tariquidar with Dox, etoposide, mitotane, and Vincristine is tested in patients with primary, recurrent, and metastatic adrenocortical cancer [49]. Regulating the protein expression levels of transporters as demonstrated that Hedgehog signaling can regulate the expression of MDR1 and ABCG2is an alternative strategy. The expression levels of ABCG2 and MDR1 are down-regulated in PC3 cells when treated with cyclo-amine (a SMO signaling element inhibitor), and targeted knockdown of ABCG2 and MDR1 expression by siRNA reverses chemoresistance [53].

2.3.3 Targeting key signaling cascades

Active anti-apoptotic and parallel inactive pro-apoptotic pathways are attracting research interest. Monoclonal antibodies targeting Notch signaling might be beneficial as targets [55, 56]. Inhibition of Notch1 can significantly reduce the CD44+CD24−/low subpopulation and reduce the incidence of brain metastases from a breast cancer cell line [57] while increased levels of β-catenin correlate with CSC tumorigenicity in colon cancer [58]. Inhibitors of Wnt signaling are also being studied [59, 60].

Monoclonal antibodies against the Wnt cascade show anti-tumor activity [61] and small-molecule Hedgehog antagonists have shown the potential to inhibit systemic metastases in mice with orthotopic xenografts from human pancreatic cancer cell lines. A reduction in ALDH-positive cells (the tumor-initiating population in pancreatic cancer) has been observed [62] and inhibiting NF-κB can suppress chemoresistance, while inhibitors targeting NF-κB can mediate anti-tumor responses and enhance the sensitivity of tumor cells to anti-cancer drugs. Co-delivery of Dox and the NF-κB inhibitor PDTC also overcomes MDR [63]. Ceramide is a secondary lipid messenger inducing the regulation of cell proliferation, apoptosis and differentiation, and acts as a pro-apoptotic molecule that activate apoptotic pathways via receptor-independent mechanisms [64]. IL-4 protect tumorigenic CD133+ CSCs in human colon carcinoma from apoptosis and therefore an anti-IL-4 antibody or IL-4Ra antagonist sensitize CSCs to chemotherapeutic drugs and cause CSCs to become apoptotic [65].

2.3.4 Targeting the tumor microenvironment

The tumor microenvironment creates a niche to protect CSCs from drug-induced apoptosis. Most mature B-cell malignancies are incurable, although evidence shows that accessory stromal cells in the tissue microenvironments of bone marrow and secondary lymphoid organs induces disease progression by promoting malignant B-cell growth, proliferation and drug resistance [66]. Tumor angiogenesis is related to CSC survival and drug resistance, and VEGF accelerates microvasculature formation and tumor growth. Thus, targeting VEGF with bevacizumab induces normal tumor vasculature and cause a disruption of the CSC niche. Treatment of mouse glioblastoma with bevacizumab induces a reduction in glioblastoma stem cells [67]. Acidic pH is characteristic of the microenvironment of solid tumors. TAT peptide is an arginine-rich peptide (YGRKKRRQRRR) that rapidly transports attached molecules into mammalian cells both *in vitro* and *in vivo*, and Lee *et al* [68] introduced TAT to a pH-sensitive cargo that was also conjugated to biotin, exposing it on its surface under slightly acidic pH values (6.5 < pH < 7.2). The pH environment together with exposed biotin increase the specificity of TAT, which elevates the concentration of Dox in the cytosol and increase the potency of Dox. Evaluated in tumor xenografts of non-small cell lung cancer A549 cells, A2780/AD-resistant breast cancer MCF-7 cells, and nasopharyngeal carcinoma cells showed a reduction in the size of all tumors [68].

References

[1] Reya T, Morrison S J, Clarke M F and Weissman I L 2001 Stem cells, cancer, and cancer stem cells *Nature* **414** 105–11

[2] Bruce W R and Van Der Gaag H 1963 A quantitative assay for the number of murine lymphoma cells capable of proliferation *in vivo Nature* **199** 79–80

[3] Bonnet D and Dick J E 1997 Human acute myeloid leukemia is organized as a hierarchy that originates from a primitive hematopoietic cell *Nature Med.* **3** 730–7

[4] Al-Hajj M, Wicha M S, Benito-Hernandez A, Morrison S J and Clarke M F 2003 Prospective identification of tumorigenic breast cancer cells *Proc. Natl Acad. Sci. USA* **100** 3983–8

[5] Chen K, Huang Y and Chen J 2013 Understanding and targeting cancer stem cells: therapeutic implications and challenges *Acta Pharmacol. Sinica* **34** 732–40

[6] Visvader J E and Lindeman G J 2008 Cancer stem cells in solid tumours: accumulating evidence and unresolved questions *Nat. Rev. Cancer* **8** 755–68

[7] Lobo N A, Shimono Y and Qian D *et al* 2007 The biology of cancer stem cells *Ann. Rev. Cell Dev. Biol.* **23** 675–99

[8] Liao W, Ye Y P and Deng Y J *et al* 2014 Metastatic cancer stem cells: from the concept to therapeutics *Am. J. Stem Cells* **3** 46–62

[9] Karsten U and Goletz S 2013 What makes cancer stem cell markers different? *Springerplus* **2** 301

[10] Sharpe B, Beresford M and Bowen R *et al* 2013 Searching for prostate cancer stem cells: markers and methods *Stem Cell Rev.* **9** 721–30

[11] Wang B, Han J, Gao Y, Xiao Z, Chen B, Wang X, Zhao W and Dai J 2007 The differentiation of rat adipose-derived stem cells into OEC-like cells on collagen scaffolds by co-culturing with OECs *Neurosci. Lett.* **421** 191–6

[12] Vainstein V, Kirnasovsky O U and Kogan Y *et al* 2012 Strategies for cancer stem cell elimination: insights from mathematical modeling *J. Theor. Biol.* **298** 32–41

[13] Han J, Fujisawa T and Husain S R *et al* 2013 Identification and characterization of cancer stem cells in human head and neck squamous cell carcinoma *BMC Cancer* **14** 1–11

[14] Horton S J and Huntly B J P 2013 Recent advances in acute myeloid leukemia stem cell biology *Haematologica* **97** 966–74

[15] Collins A T, Berry P A and Hyde C *et al* 2005 Prospective identification of tumorigenic prostate cancer stem cells *Cancer Res.* **65** 10946–51

[16] Li C W, Heidt D G and Dalerba P *et al* 2007 Identification of pancreatic cancer stem cells *Cancer Res.* **67** 1030–7

[17] Okamoto A, Chikamatsu K and Sakakura K *et al* 2009 Expansion and characterization of cancer stem-like cells in squamous cell carcinoma of the head and neck *Oral Oncology* **45** 633–9

[18] Eramo A, Lotti F and Sette G *et al* 2008 Identification and expansion of the tumorigenic lung cancer stem cell population *Cell Death Differ.* **15** 504–14

[19] Kim H, Choi G H and Na D C *et al* 2009 Human hepatocellular carcinomas with 'Stemness'-related marker expression: keratin 19 expression and a poor prognosis *Hepatology* **54** 1707–17

[20] Bussolati B, Bruno S and Grange C *et al* 2008 Identification of a tumor-initiating stem cell population in human renal carcinomas *FASEB J.* **22** 3696–705

[21] Tang D G 2012 Understanding cancer stem cell heterogeneity and plasticity *Cell Res.* **22** 457–72

[22] Campbell L L and Polyak K 2007 Breast tumor heterogeneity: cancer stem cells or clonal evolution? *Cell Cycle* **6** 2332–8

[23] Kreso A *et al* 2013 Variable clonal repopulation dynamics influence chemotherapy response in colorectal cancer *Science* **339** 543–8

[24] Anderson K *et al* 2011 Genetic variegation of clonal architecture and propagating cells in leukaemia *Nature* **469** 356–61

[25] Notta F *et al* 2011 Evolution of human BCR-ABL1 lymphoblastic leukaemia-initiating cells *Nature* **469** 362–7

[26] Heppner G H and Miller B E 1983 Tumor heterogeneity: biological implications and therapeutic consequences *Cancer Metastasis Rev.* **2** 5–23

[27] Southam C M and Brunschwig A 1961 Quantitative studies of autotransplantation of human cancer *Cancer* **14** 971–8

[28] Furth J and Kahn M C 1937 The transmission of leukemia in mice with a single cell *Am. J. Cancer* **31** 276–82

[29] Hewitt H B 1958 Studies of the dissemination and quantitative transplantation of a lymphocytic leukaemia of CBA mice *Br. J. Cancer* **12** 378–401

[30] Hamburger A W and Salmon S E 1977 Primary bioassay of human tumor stem cells *Science* **197** 461–3

[31] Cho K, Wang X, Nie S, Chen Z and Shin D M 2008 Therapeutic nanoparticles for drug Delivery in cancer *Clin. Cancer Res.* **14** 1310–6

[32] LaBarge M A 2010 The difficulty of targeting cancer stem cell niches *Clin. Cancer Res.* **16** 3121–9

[33] Lacerda L, Pusztai L and Woodward W A 2010 The role of tumor initiating cells in drug resistance of breast cancer: implications for future therapeutic approaches *Drug Resist. Updates* **13** 99–108

[34] Wang Z *et al* 2010 Targeting miRNAs involved in cancer stem cell and EMT regulation: an emerging concept in overcoming drug resistance *Drug Resist. Updates* **13** 109–18

[35] Liu J *et al* 2009 Biorecognition and subcellular trafficking of HPMA copolymer-anti-PSMA antibody conjugates by prostate cancer cells *Mol. Pharm.* **6** 959–70

[36] Curiel T J 2012 Immunotherapy: a useful strategy to help combat multidrug resistance *Drug Resist. Updates* **15** 106–13

[37] Jin L, Hope K J, Zhai Q, Smadja-Joffe F and Dick J E 2006 Targeting of CD44 eradicates human acute myeloid leukemic stem cells *Nat. Med.* **12** 1167–74

[38] Jin L, Hope K J, Zhai Q, Smadja-Joffe F and Dick J E 2006 Targeting of CD44 eradicates human acute myeloid leukemic stem cells *Nat. Med.* **12** 1167–74

[39] Kikushige Y *et al* 2010 TIM-3 is a promising target to selectively kill acute myeloid leukemia stem cells *Cell Stem Cell* **7** 708–17

[40] Chao M P *et al* 2011 Therapeutic antibody targeting of CD47 eliminates human acute lymphoblastic leukemia *Cancer Res.* **71** 1374–84

[41] Alkilany A M, Thompson L B, Boulos S P, Sisco P N and Murphy C J 2012 Gold nanorods: their potential for photothermal therapeutics and drug delivery, tempered by the complexity of their biological interactions *Adv. Drug Deliv. Rev.* **64** 190–9

[42] Gundlach C W *et al* 2011 Synthesis and evaluation of an anti-MLC1×anti-CD90 bispecific antibody for targeting and retaining bone-marrow-derived multipotent stromal cells in infarcted myocardium *Bioconj. Chem* **22** 1706–14

[43] Lin E H *et al* 2007 Elevated circulating endothelial progenitor marker CD133 messenger RNA levels predict colon cancer recurrence *Cancer* **110** 534–42

[44] Mehra N *et al* 2006 Progenitor marker CD133 mRNA Is elevated in peripheral blood of cancer patients with bone metastases *Clin. Cancer Res.* **12** 4859–66

[45] Wang C H, Chiou S H, Chou C P, Chen Y C, Huang Y J and Peng C A 2011 Photothermolysis of glioblastoma stem-like cells targeted by carbon nanotubes conjugated with CD133 monoclonal antibody *Nanomedicine* **7** 69–79

[46] Brescia P, Ortensi B, Fornasari L, Levi D, Broggi G and Pelicci G 2013 CD133 is essential for glioblastoma stem cell maintenance *Stem Cells* **31** 857–69

[47] Broxterman H J, Gotink K J and Verheul H M W 2009 Understanding the causes of multidrug resistance in cancer: a comparison of doxorubicin and sunitinib *Drug Resist. Updates* **12** 114–26

[48] Shapira A, Livney Y D, Broxterman H J and Assaraf Y G 2011 Nanomedicine for targeted cancer therapy: towards the overcoming of drug resistance *Drug Resist. Updates* **14** 150–63

[49] Patil Y, Sadhukha T, Ma L and Panyam J 2009 Nanoparticle-mediated simultaneous and targeted delivery of Paclitaxel and tariquidar overcomes tumor drug resistance *J. Control. Release* **136** 21–9

[50] Ritchie T K, Kwon H and Atkins W M 2011 Conformational analysis of Human ATP-binding cassette transporter ABCB1 in lipid nanodiscs and inhibition by the antibodies MRK16 and UIC2 *J. Biol. Chem.* **286** 39489–96

[51] Tsuruo T, Iida H, Tsukagoshi S and Sakurai Y 1981 Overcoming of Vincristine resistance in P388 leukemia *in vivo* and *in vitro* through enhanced cytotoxicity of Vincristine and vinblastine by Verapamil *Cancer Res.* **41** 1967–72

[52] Saeki T *et al* 2007 Dofequidar fumarate (MS-209) in combination with cyclophosphamide, doxorubicin, and fluorouracil for patients with advanced or recurrent breast cancer *J. Clin. Oncol.* **25** 411–7

[53] Minderman H, O'Loughlin K L, Pendyala L and Baer M R 2004 VX-710 (Biricodar) increases drug retention and enhances chemosensitivity in resistant cells overexpressing P-glycoprotein, multidrug resistance protein, and breast cancer resistance protein *Clin. Cancer Res.* **10** 1826–34

[54] Sims-Mourtada J, Izzo J G, Ajani J and Chao K S C 2007 Sonic hedgehog promotes multiple drug resistance by regulation of drug transport *Oncogene* **26** 5674–9

[55] Fischer M *et al* 2011 Anti-DLL4 inhibits growth and reduces tumor-initiating cell frequency in colorectal tumors with oncogenic KRAS mutations *Cancer Res.* **71** 1520–5

[56] Li K *et al* 2008 Modulation of notch signaling by antibodies specific for the extracellular negative regulatory region of NOTCH3 *J. Biol. Chem.* **283** 8046–54

[57] McGowan P M *et al* 2011 Notch1 inhibition alters the CD44hi/CD24lo population and reduces the formation of brain metastases from breast cancer *Mol. Cancer Res.* **9** 834–44

[58] Vermeulen L *et al* 2010 Wnt activity defines colon cancer stem cells and is regulated by the microenvironment *Nat. Cell. Biol.* **12** 468–76

[59] Chen B *et al* 2009 Small molecule-mediated disruption of Wnt-dependent signaling in tissue regeneration and cancer *Nat. Chem. Biol.* **5** 100–7

[60] Fujii N *et al* 2007 An antagonist of dishevelled protein–protein interaction suppresses β-caten independent tumor cell growth *Cancer Res.* **67** 573–9

[61] He B *et al* 2005 Blockade of Wnt-1 signaling induces apoptosis in human colorectal cancer cells containing downstream mutations *Oncogene* **24** 3054–8

[62] Feldmann G *et al* 2008 An orally bioavailable small-molecule inhibitor of hedgehog signaling inhibits tumor initiation and metastasis in pancreatic cancer *Mol. Cancer Ther.* **7** 2725–35

[63] Fan L *et al* 2010 Co-delivery of PDTC and doxorubicin by multifunctional micellar nanoparticles to achieve active targeted drug delivery and overcome multidrug resistance *Biomaterials* **31** 5634–42

[64] Devalapally H, Duan Z, Seiden M V and Amiji M M 2008 Modulation of drug resistance in ovarian adenocarcinoma by enhancing intracellular ceramide using tamoxifen-loaded biodegradable polymeric nanoparticles *Clin. Cancer Res.* **14** 3193–203

[65] Todaro M *et al* 2007 Colon cancer stem cells dictate tumor growth and resist cell death by production of interleukin-4 *Cell Stem Cell* **1** 389–402

[66] Konopleva M, Tabe Y, Zeng Z and Andreeff M 2009 Therapeutic targeting of micro-environmental interactions in leukemia: mechanisms and approaches *Drug Resist. Updates* **12** 103–13

[67] Burkhardt J K *et al* 2012 Orthotopic glioblastoma stem-like cell xenograft model in mice to evaluate intra-arterial delivery of bevacizumab: from bedside to bench *J. Clin. Neurosci.* **19** 1568–72

[68] Lee E S, Gao Z, Kim D, Park K, Kwon I C and Bae Y H 2008 Super pH-sensitive multifunctional polymeric micelle for tumor pH specific TAT exposure and multidrug resistance *J. Control. Release* **129** 228–36

IOP Concise Physics

Photomedicine and Stem Cells
The Janus face of photodynamic therapy (PDT) to kill cancer stem cells, and photobiomodulation
(PBM) to stimulate normal stem cells
Heidi Abrahamse and Michael R Hamblin

Chapter 3

PDT and cancer stem cells

3.1 PDT and cancer treatment

Photodynamic therapy (PDT) is a highly specific anticancer treatment method and is used for a variety of cancers, predominantly for recurrent cancers that no longer respond to conventional anti-cancer therapies. It is a treatment modality comprising photosensitizers (PSs), low-intensity laser irradiation (photobiomodulation (PBM)) and oxygen. A PS is a molecule that localizes in a specific target cell and becomes activated when exposed to a light source. When exposed to a specific wavelength the molecule becomes excited and then returns to the ground state. The energy released mediates selective cell killing in two ways. In the first reaction (type I reaction), the PS reacts with biomolecules through a hydrogen atom (electron) transfer to form radicals, which react with molecular oxygen to generate reactive oxygen species (ROS). This in turn produces oxidative stress and results in cell death. In the second type (type II), the most common reaction, energy is transferred directly to oxygen in the cell to form a singlet oxygen (a subset of ROS) which then oxidizes various substrates resulting in cell killing. PDT has been shown to be an effective treatment modality in cancers of the lung, skin, breast, head and neck, digestive tract, pancreas, liver, bladder, ovary, prostate, and brain.

3.2 PDT and CSCs

CSCs are thought to be resistant to conventional cancer therapies, including PDT, but resistance to PDT depends on a variety of factors. These include: PS uptake, and localization, alterations in the expression and function of key molecules involved in PS transport, the loss of apoptosis and autophagy, induction of antioxidant defenses, the induction of heat shock protein changes in cytoskeleton and adhesion, the induction of cyclooxygenases, the production of nitric oxide, and hypoxia. This being said, there has been little evidence that completely excludes PDT as a CSC treatment modality. Studies by Yu and Yu [1] reported that 5-aminolevulinic acid, used in PDT treatment of head and neck CSCs, impaired their tumor-initiating and chemoresistance potential. Another study conducted by Wei in 2014 [2], used protoporphyrin IX PDT to treat colorectal CSCs and also reported the efficacy of PDT in their treatment. Other studies have used modified PDT techniques to overcome the reduced oxygen levels which limit the efficacy of PDT in the tumor microenvironments housing CSCs by using surfactant–polymer nanoparticles to increase production of ROS. Hyperthermia (HPT) in combination with PDT has also been explored and studies on glioblastoma have shown that the combined HPT and PDT approach is quite effective in treating this type of cancer. Additionally, the development of functional nanocarriers can efficiently overcome issues regarding multiple drug delivery and multi-drug resistance (MDR) [3].

3.3 ABC transporters and ABCG2

Adenosine triphosphate (ATP)-binding cassette (ABC) transporters function as a defense system against toxic molecules absorbed from the environment, using energy from ATP hydrolysis to pump xenobiotics out of cells [4]. ABC transporters are a protein superfamily, which is ubiquitous and highly conserved in various lifeforms in nature. They are characterized by a consensus ATP-binding region of approximately 90–110 amino acids, including the Walker A and B motifs, between which lies the dodecapeptide linker region (Walker C region). The transporters also usually contain transmembrane (TM) domains, which generally consist of six TM helices that confer substrate specificity. There are 48 known ABC transporters, which are expressed in humans. Though the development of ABC transporters was a major advantage in the evolution of mankind, they can also function as a hindrance to treatment (in particular cancer chemotherapy) or as a causative factor for various diseases. These efflux pumps are often quite indiscriminate, and therefore have a wide range of substrates including many therapeutic drugs used to treat diseases. In addition, there are at least 13 genetic diseases, including adrenoleukodystrophy, Tangier disease, and Dubin–Johnson syndrome, which have been attributed to defects in ABC transporters [4]. ABC genes are divided into seven distinct subfamilies: ABC1, MDR/TAP, MRP, ALD, OABP, GCN20, and White [5].

ABCG2 (ABC sub-family G member 2) is a protein that in humans is encoded by the ABCG2 gene. The ABCG2 gene is located on chromosome 4q22, spans more than 66 kilobases, and consists of 16 exons ranging from 60 to 532 base pairs [6]. The membrane-associated protein encoded by this gene is included in the superfamily of ABC transporters. ABCG2 is a member of the White subfamily. ABCG2 is sometimes referred to as the 'breast cancer resistance protein' (BCRP), and functions as a xenobiotic transporter which may play a role in MDR to chemotherapeutic agents, including mitoxantrone and camptothecin analogs. Early observations of significant ABCG2-mediated resistance to anthracyclines were subsequently attributed to mutations encountered *in vitro* but not in nature or the clinic. Significant expression of this protein has been observed in the placenta [7], and it has been shown to have a role in protecting the fetus from xenobiotics in the maternal circulation [8].

ABCG2 has also been shown to play protective roles in blocking absorption of potentially toxic foreign molecules at the apical membrane of the intestine, and at the blood–testis barrier, the blood–brain barrier, and the membranes of hemato-poietic progenitor and other stem cells [9]. At the apical membranes of the liver and kidney, it enhances excretion of xenobiotics [10]. In the lactating mammary gland, it has a role on excreting vitamins such as riboflavin and biotin into breast milk [11].

3.4 ABCG2 in cancer stem cells

ABCG2 is widely expressed in normal stem cells, and present opinion considers that ABCG2 plays an important role in promoting stem cell proliferation and the maintenance of the stem cell phenotype. ABCG2 has also been found to confer the side population phenotype in CSCs. Moreover, ABCG2 expression in tumors may contribute to their formation and progression [12]. Taking into account that the SP phenotype is mainly mediated by ABCG2 and the conserved expression of ABCG2 in stem cells, it is likely that ABCG2 may serve as a biomarker of CSCs. Since ABCG2 functions as a high-capacity efflux transporter with a wide range of substrates, including various chemotherapy drugs, it has been shown to participate in the MDR of tumors, which is responsible for the eventual failure of many (if not most) common chemotherapy regimens [13]. CSCs have also been held responsible for the emergence of multi-drug chemotherapy resistance and eventual cancer recurrence. The correlation between the occurrence of the side population and chemoresistance suggest a close link between ABCG2 and CSCs. ABCG2+ tumor cells may hence represent a unique population of CSCs.

Elevated expression of ABCG2 has been observed in a number of putative CSCs from retinoblastoma [14], lung [15], liver [16] and pancreatic cancer [17]. In addition, ABCG2 and CD133, the widely identified CSC markers are co-expressed in melanoma [18] and pancreatic carcinoma cell lines [19]. ABCG2+ populations showed evidence for self-renewal, generating both ABCG2+ and ABCG2– prog-enies during subculture, and a higher proliferative activity. Moreover, other progenitor cell markers including cytokeratin 19 and alphafetoprotein were mainly expressed in ABCG2+ subpopulations [20].

3.5 ABCG2 and PS efflux

The first studies that questioned whether efflux pumps played a role in the effectiveness of PDT for cancer date from 1989. Cowled and Forbes [21] had studied several vasoactive drugs in combination with hematoporphyrin derivative (HPD), with both injected together into tumor-bearing mice 24 h before light delivery, or with the HPD injected 24 h and the vasoactive drugs 2 h before light. Noradrenaline, propranolol, hydralazine, and phenoxybenzamine inhibited PDT effects on tumors if injected 2 h before irradiation. This inhibition was associated with reduced uptake of HPD into tumors. There was no inhibition if irradiation was delayed until 24 h after administration of the vasoactive drug, presumably because HPD uptake continued after the drugs had ceased to affect the vasculature. Verapamil enhanced photodynamic destruction of tumors when administered concurrently with HPD (24 h before light) and the enhancement was associated with increased uptake of HPD into tumors. Verapamil neither increased uptake of HPD nor enhanced PDT killing of cells *in vitro*. When Verapamil was administered after irradiation, regrowth of tumors was inhibited. Other calcium channel blocking agents, diltiazem and nifedipine, had no effect on the uptake of HPD or the inhibition of the regrowth of tumors after PDT. Gossner *et al* [22] carried out a similar study, injecting Verapamil concurrently with HPD 24 h before light delivery in two different mouse tumor models.

The role of ABCG2 in mediating the efflux of dietary tetrapyrrole derivatives was discovered by accident in 2003. ABCG2(−/−) mice were constructed by Jonker *et al* in order to study whether anti-cancer drugs such as topotecan could be administered orally in the absence of ABCG2 [23]. The oral availability of topotecan was increased six-fold in ABCG2(−/−) mice, indicating that intestinal ABCG2 limits the absorption of topotecan from the gut. The ABCG2(−/−) mice had not displayed signs of any phenotypical abnormality, until a few animals suddenly developed severe necrotic ear lesions. Only mice housed in cages on the top shelf of the vivarium racks, closest to the room lighting, developed these lesions, suggesting some form of phototoxicity. Further studies showed that all ABCG2(−/−) mice were able to develop ear lesions when exposed to standard fluorescent light, but only when fed with mouse chow containing alfalfa leaf concentrate. Phototoxic ear lesions developed one week after feeding with this chow, and in some cases lesions also appeared on the tail, snout, and rims of the eyes. Phototoxicity was never observed in wild-type mice. Erythrocyte levels of the heme precursor, protoporphyrin IX, which is structurally related to pheophorbide-a, were increased ten-fold. Transplantation with wild-type mouse bone marrow cured the protoporphyria and reduced the phototoxicity in the mice. It was pointed out that cattle that have been fed with alfalfa at various times have been reported to suffer outbreaks of photosensitivity.

Given the foregoing information, it was not at all surprising when a number of PS that are used for PDT of cancer turned out to be substrates of ABCG2 [24]. Robey *et al* reported that pheophorbide-a, pyropheophorbide-a methyl ester (MPPa), chlorin (e6), protoporphyrin IX, and hematoporphyrin IX were all substrates of ABCG2, while meso-tetra(3-hydroxyphenyl)porphyrin and meso-tetra(3-hydroxyphenyl) chlorin were not substrates [25]. Liu *et al* [26] described energy-dependent efflux of 2-(1-hexyloxethyl)-2-devinyl pyropheophorbide-a (HPPH, Photochlor), endogenous protoporphyrin IX synthesized from 5-aminolevulinic acid, and benzoporphyrin derivative monoacid ring A (BPD-MA, Verteporfin) in ABCG2+ cell lines, but the first-generation PS porfimer sodium (Photofrin) and a novel derivative of HPPH conjugated to galactose were only minimally transported. Morgan *et al* [27] tested a series of conjugates of PS derived from chlorophyll-a (pyropheophorbides and purpurinimides) with different groups attached at different positions on the tetrapyrrole macrocycle to examine whether a change in affinity for the ABCG2 pump occurred. Carbohydrate groups conjugated at positions 8, 12, 13, and 17 but not at position 3 abrogated ABCG2 affinity regardless of structure or linking moiety. At position 3, affinity was retained with the addition of iodobenzene, alkyl chains, and

monosaccharides, but not with disaccharides. This suggests that structural characteristics at position 3 may offer important contributions to requirements for binding to
ABCG2. The substrate PS HPPH (2-[1-hexyloxyethyl]-2-devinyl pyropheophorbide-
a) preserved the SP fraction of tumor cells, but the non-substrate PS HPPH-Gal (a
galactose conjugate of HPPH) was able to eradicate them by PDT.

3.6 Efflux pump inhibition and PDT

Since it was discovered that many PSs are substrates of ABCG2, and that this efflux
was particularly important in the case of CSCs, several workers have investigated
strategies to overcome the ABCG2-mediated efflux of PSs. One of the first reports was
from Liu *et al* [26] who tested the tyrosine kinase inhibitor (TKI) imatinib mesylate
(Gleevec) on HPPH accumulation and *in vitro* and *in vivo* PDT efficacy. Imatinib
mesylate increased accumulation of HPPH, PpIX, and BPD-MA from 1.3- to 6-fold in
ABCG2+ cells, but not in ABCG2− cells, and enhanced PDT efficacy both *in vitro* and
in vivo. Pan *et al* [28] used a different ABCG2 inhibitor, fumitremorgin C (FTC) to test
potentiation of PDT mediated by MPPa in four human glioma cell lines (U87, A172,
SHG-44, and U251). The intracellular MPPa and ROS in A172 receiving MPPa–PDT
significantly increased after using the ABCG2 inhibitor. Both cell viability and
apoptosis in A172 cells undergoing MPPa–PDT were significantly improved with
FTC.

3.7 Evidence of effect of PBM on CSCs

Work performed in our laboratory, Laser Research Centre, Faculty of Health
Sciences, University of Johannesburg, explored the possibility of outcomes using
PBM on lung CSCs. We investigated the effects of PBM treatment of lung CSCs.
Prominin 1 (CD133), a pentaspan TM glycoprotein which is an antigenic structure
usually found on stem cell surfaces [29] was used in this study for positive
identification and isolation of CD133+ lung CSCs. Results of this study showed
positive isolation, confirming research by Tirino [30] that CSCs can be identified and
isolated using their antigenic markers. CD133 has previously been used to identify and
isolate different stem cells and CSCs [31]. The cell lines used as a control for antigenic
identification of CD133 were Caco-2, which is a colorectal carcinoma cell line positive
for the antigen CD133 [31], and SKUT-1 human uterus leiomyosarcoma, which does
not express the surface marker CD133 and was used as the negative control cell line.

Analysis of lung CSC morphology post irradiation revealed that cells receiving no
irradiation maintained their cell morphology, showing an increase in cell density
over time due to normal cell growth. Similar results were seen for irradiated test
samples receiving LF-PBM. Stimulation of viability and proliferation were found,
seen as an increase in cell density, when examining test samples receiving PBM of 5
and 10 J cm^{-2} with a wavelength of 636 nm and 825 nm [32]. Findings are confirmed
by studies also suggesting that PBM can stimulate cells with a wavelength of 636–
825 nm and low fluence of 5–10 J cm^{-2} [33, 34]. Morphological discrepancies,
indicated by cellular retraction and cell rounding, were found when using HF-PBM
of 40 J cm^{-2}, which is suggestive of bio-inhibition that was countered over time due

to cellular recovery followed by normal cycle maintenance. This characteristic is also found in SCs where the cells stay quiescent until regeneration of cells stimulate proliferation via a specific signal [35], allowing the irradiated CSCs to proliferate at a similar rate.

Observations seen from viability and proliferation analysis revealed that different wavelengths at different fluences have either a biostimulatory or bio-inhibitory effect on isolated lung CSCs. The greatest effect was found when using LF-PBM of 10 J cm^{-2} and a wavelength of 825 nm. LF-PBM showed stimulation of cells with a wavelength of 636–825 nm and fluence of 5–10 J cm^{-2} concurring with previous researchers who have found similar results [33, 34, 36]. CSCs oppose cell death by self-renewal [37], which could be seen when treating cells with increased energy levels of 20 J cm^{-2}, when compared to non-cancerous cell lines which have died during exposure to increased energy levels [33]. These findings suggest that irradiation between 5 and 20 J cm^{-2} with wavelengths ranging from 636–060 nm has few damaging effects, but rather a stimulatory effect on the CSCs.

AlamarBlue™ was used to measure the amount of proliferating cells (figure 3.1). Measurement was read as an absorbance value with an increase in proliferation over several hours observed in all samples. When comparing non-irradiated samples to treated samples a statistically significant difference was observed in all fluences used on the cells during their proliferation time of 24–72 h post irradiation ($p < 0.001$).

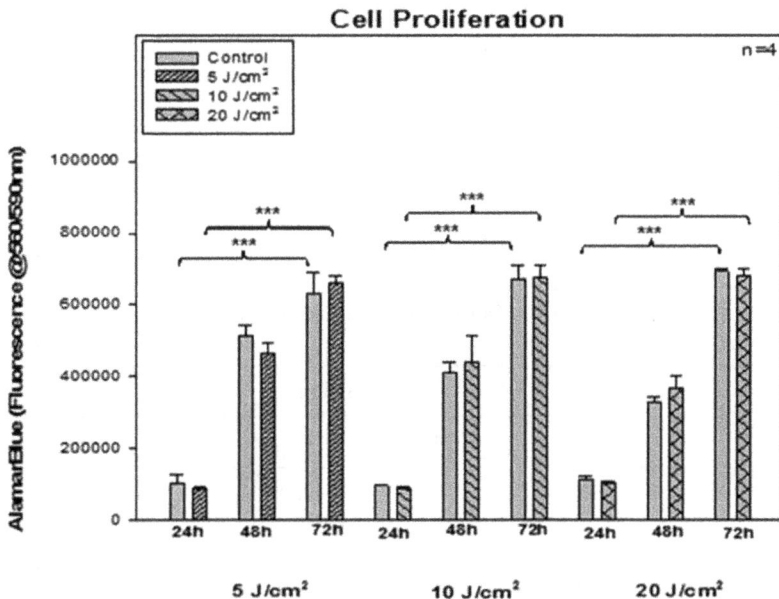

Figure 3.1. The AlamarBlue assay was used to measure cell proliferation. Increase in proliferation is seen in all the samples as time elapses from 24 to 72 h. The non-irradiated cell's proliferation time is compared to the treated cell's proliferation time. Statistical significant differences are expressed as $p < 0.05$ (*), $p < 0.01$ (**), and $p < 0.001$ (***).

This study indicates the possible detrimental effect that PBM may have when used as a biostimulatory therapy on the underlying tissue CSCs when considering the proliferation and viability induced using visible wavelengths.

Significant bio-inhibition was seen in viability and proliferation when irradiating samples with HF-PBM at 40 J cm^{-2}. These findings are supported by previous research stating that parameters such as wavelength, fluence, and intensity play an important role in PBM of cellular metabolism [33, 36, 38, 39]. Findings also revealed that PBM with a wavelength of 1060 nm and a fluence of 5–20 J cm^{-2} had little effect on the treated samples, showing no stimulation or inhibition of cellular metabolism. This indicates that CSC photobiostimulation is only achieved when using irradiation having wavelengths in the visible red-light spectrum ranging between 636 and 825 nm. Infrared light (1060 nm) has little to no effect on CSC metabolism using LF-PBM [32].

Cytotoxicity analysis revealed membrane damage of isolated lung CSCs by measuring the release of lactate dehydrogenase (LDH) in the medium. Cells treated with low fluence photobiomodulation (LF-PBM) (5–20 J cm^{-2}) using all respective wavelengths did not show any membrane damage as no significant amounts of LDH were released. Membrane damage was seen when irradiating cells using high fluence photobiomodulation (HF-PBM) of 40 J cm^{-2} showing high levels of LDH release.

The level of LDH activity was determined using the CytoTox96® non-radioactive cytotoxicity assay. The amount of LDH released into the culture medium was measured and is indicative of the level of cell damage related to membrane damage of the isolated lung CSCs.

LDH released from the cytosol was quantified by spectroscopy. Treated cells, when compared to their controls, released notable but insignificant amounts of LDH at all wavelengths using fluences of 5–20 J cm^{-2}. An increase in cell damage was observed with treatment of 40 J cm^{-2} at wavelengths of 636 nm, 825 nm, and 1060 nm, and when compared to the controls and other fluences was significant having a p-value of <0.05 (tables 3.1–3.3) [40].

Table 3.1. LDH cytotoxicity of isolated CD133 lung CSCs measured at 636 nm. Test samples received irradiation of 5–40 J cm^{-2}. A significant amount of toxicity was only observed when comparing test samples to their controls at 40 J cm^{-2} ($p < 0.05$ (*)). Other comparisons did not yield any significant LDH cytotoxicity ($p < 0.01$ or $p < 0.001$).

		636 nm					
		24 h		48 h		72 h	
5 J cm^{-2}	Control	0.3470	±0	0.3950	±0	0.3900	±0
	Test	0.3440	±0	0.3883	±0	0.4008	±0.01
10 J cm^{-2}	Control	0.3980	±0	0.4360	±0	0.4470	±0
	Test	0.4263	±0.01	0.4540	±0.01	0.4513	±0.01
20 J cm^{-2}	Control	0.5630	±0	0.6080	±0	0.6490	±0
	Test	0.6183	±0.02	0.6123	±0	0.6535	±0.02
40 J cm^{-2}	Control	0.6780	±0	0.7100	±0	0.6580	±0
	Test	1.1998	*±0.11	0.9570	*±0.04	0.8673	*0.01

Table 3.2. LDH cytotoxicity of isolated CD133 lung CSCs measured at 825 nm. The test samples received irradiation of 5–40 J cm^{-2}. When comparing test samples to their non-irradiated controls only 40 J cm^{-2} showed significant LDH release ($p < 0.05$ (*)). Other comparisons did not yield any significant LDH cytotoxicity ($p < 0.01$ or $p < 0.001$).

		825 nm					
		24 h		48 h		72 h	
5 J cm^{-2}	Control	0.3070	±0	0.3520	±0	0.4020	±0
	Test	0.3563	±0.02	0.3675	±0.01	0.4363	±0.02
10 J cm^{-2}	Control	0.4210	±0	0.4330	±0	0.4410	±0
	Test	0.4265	±0.01	0.4258	±0.01	0.4398	±0
20 J cm^{-2}	Control	0.5950	±0	0.6000	±0	0.6310	±0
	Test	0.6293	±0.01	0.6413	±0.01	0.6473	±0.01
40 J cm^{-2}	Control	0.6910	±0	0.6720	±0	0.7020	±0
	Test	1.3306	*±0.07	1.0738	*±0.06	1.0405	*±0.01

Table 3.3. LDH cytotoxicity of isolated CD133 lung CSCs measured at 1060 nm. The irradiation fluence ranged from 5–40 J cm^{-2}. Controls received no irradiation. When comparing tests to their respective controls only 40 J cm^{-2} indicated a significant amount of cytotoxicity ($p < 0.05$ (*)). Changes seen in other fluences did not have a significant amount of LDH leakage, ($p < 0.01$ or $p < 0.001$).

		1060 nm					
		24h		48h		72h	
5 J cm^{-2}	Control	0.3280	±0	0.3710	±0	0.3490	±0
	Test	0.3348	±0	0.3855	±0	0.3963	±0.02
10 J cm^{-2}	Control	0.4090	±0	0.4290	±0	0.4180	±0
	Test	0.4083	±0	0.4385	±0.01	0.4430	±0.01
20 J cm^{-2}	Control	0.4890	±0	0.5630	±0	0.5810	±0
	Test	0.5310	±0.01	0.5840	±0.01	0.5935	±0
40 J cm^{-2}	Control	0.6240	±0	0.6670	±0	0.6580	±0
	Test	1.1878	*±0.04	0.9360	*±0.01	0.8330	*±0.04

The membrane damage found is also a direct indication of cellular death. Cell death was measured using Annexin V PI. Little to no cell death was seen when irradiating the cells at fluences of 5–20 J cm^{-2} with wavelengths of 636–1060 nm as findings show no apoptosis. Cellular apoptosis was seen when using HF-PBM of 40 J cm^{-2} and wavelengths of 636, 825, and 1060 nm. Results revealing necrotic cell death at all irradiation parameters were due to cells going into late apoptosis and eventually dying [32]. These results are also supported by reports indicating that light is absorbed by chromophores (porphyrins or cytochromes) located in the intercellular organelles such as the plasma membrane, mitochondria, or lysomes. Stimulation through LF-PBM had no bio-inhibitory effect with no cell damage that could lead to the cell death observed. HF-PBM had a bio-inhibitory effect, revealing membrane damage and activated cell death due to ROS production [41]. Findings indicate that HF-PBM can induce cell death through apoptosis, although the exact mechanism is not well understood.

Study findings revealed that LF-PBM and HF-PBM of isolated lung CSCs can induce biostimulation and bio-inhibition, respectively. This was indicated by results indicating an increase in ATP production after exposure to irradiation at 5–10 J cm^{-2} with wavelengths of 636 nm and 825 nm. A decrease in ATP production was observed after exposure to HF-PBM (40 J cm^{-2}). Proliferation was affected in a similar manner compared to ATP viability, as ATP is a direct indication of metabolic activity. These findings suggest that light stimulation through LF-PBM and HF-PBM can either cause an increase or a decrease in ATP production from the mitochondrion [32].

The decrease in ATP production revealed by viability and proliferation results indicate that the cell cycle of mitosis has been interrupted, along with the insufficient energy sources produced by the mitochondrion after light absorption. This is confirmed by Ocana-Quero *et al* [42], who demonstrated replication inhibition after high dosages of PBM along with DNA damage leading to a negative effect on cell proliferation.

Results from this study [32] are confirmed by reports suggesting that light is absorbed by chromophores (porphyrins or cytochromes) located in the intercellular organelles such as the plasma membrane, mitochondria, or lysomes, having different effects depending on the wavelength and fluence used [41].

It has been hypothesized that CSCs are a contributing factor to tumor initiation, metastasis, and cancer relapse [43], and this raises concerns, not only when considering current cancer treatments but also for PBM when used as a therapeutic model. Contemporary cancer treatments have limited prognosis with diminished abilities and produce adverse side effects. Alternative therapies need investigation [44] and focus needs to be turned to not only eradicating the tumor bulk, but also to include CSCs when developing a cancer treatment. When considering the CSC characteristics, PBM should be used with caution. This is due to the use of PBM for its desired effect of increased cell proliferation and viability during wound healing or stem cell regeneration [34] that can be harmful when considering treatment where CSCs are involved.

CD133-positive cells displayed higher ability of self-renewal, tumor initiation, and drug resistance [45]. Other studies also showed repression of cell differentiation and accelerated cell proliferation, anchorage-independent colony formation, and *in vivo* tumor formation using CD133-positive neuroblastoma (NB) tumor samples [46]. These specific abilities, also found in CD133-positive lung CSCs all lead to lung cancer having a high mortality rate and chance for relapse [32].

In the cancer field it has become apparent that targeted therapy has become a promising outcome and it is seen that apoptosis-triggering will play a crucial role. The focus is shifted towards achieving tumor cell specificity and exploiting the mechanism by which apoptosis is facilitated [47]. Obtaining information regarding the mechanism behind apoptotic induction of HF-PBM is necessary for the clinical application of low-power laser therapies. It can deliver apoptotic stimulatory triggers in cancer [47]. PDT has been shown to induce membrane damage and apoptosis in lung cancer leading to cell death [44], that can lead to tumor reduction and eradication.

Considering the findings from this study, future work should include identifying the effectiveness of PDT on lung CSCs, as well as the cellular death mechanisms that are activated, and whether they differ in lung cancer cells and so could possibly allow for cancer recovery when this form of chemotherapy is used [32].

Similarly, the biomodulative effect of PBM was evaluated on breast CSCs and breast cancer cells after treatment using fluencies of 5, 10, 20, 40, and 200 J cm^{-2} at 636, 825, and 1060 nm. Cell adhesion protein CD44 has been used in recent years for breast CSC identification and isolation [48]. Breast CSCs could be positively isolated from the entire cell population of the heterogenic MCF-7 cell line based on their expression of the CD44 antigenic cell surface marker. Negative expression of CD44 was observed in breast cancer cells that remained after isolation of CSCs. These cells served as the negative control throughout this study. These findings confirmed intratumoral heterogeneity—which consists of the simultaneous presence of more than one type of cell within a single neoplasm—in breast cancer [49].

Following PBM treatment using all five fluencies and all three wavelengths, no morphological changes were observed in either the breast cancer cells or breast CSCs. However, regular cell growth could be predicted by an increase in cell density in the untreated control and experimental groups that received LF-PBM (5 and 10 J cm^{-2}) at 636, 825, and 1060 nm. This was not the case in the experimental groups that were exposed to HF-PBM (200 J cm^{-2}) at 636 and 825 nm, in which obvious rounding and shrinkage, characteristic of programmed cell death, could be observed in some of the cells.

The effectiveness of PBM to induce a biostimulatory response in non-cancerous and cancerous cells has been elucidated in numerous studies [50]. Analysis of breast cancer cells and breast CSCs after treatment with LF-PBM has revealed an increase in their proliferation. In breast cancer cells, an increase in cell proliferation was observed using all three wavelengths when applying light energy of 5–40 J cm^{-2}. However, the most significant increase was seen at 825 nm using 5–10 J cm^{-2}. On the other hand, breast CSCs showed an increase in their proliferation at 636 nm and 825 nm only when applying light energy of 10 J cm^{-2}. This indicates that the breast cancer cells and breast CSCs react differently when exposed to the same treatment. The percentage of viable cells significantly increased in breast cancer cells when applying light energy of 5 J cm^{-2} and 10 J cm^{-2} at all three wavelengths. Breast CSCs showed a significant increase in their viability when using 10 J cm^{-2} at 636 nm and 5 J cm^{-2} at 825 nm. No significant increase was observed at 1060 nm. Overall, LF-PBM has an increasing effect on breast cancer cells and breast CSC proliferation and viability.

Findings suggest that HF-PBM could have bio-inhibitory effect on cancer cells [40]. However, the outcome of treatment using HF-PBM strongly differs from one cell type to another. While a light dose of 40 J cm^{-2} happened to be enough to cause cell death in lung CSCs, that was not the case for breast cancer [40]. In this study, HF-PBM of 200 J cm^{-2} at 636 and 825 nm was able to induce a decrease in proliferation and the percentage of viable cells in both breast cancer cells and breast CSCs.

Post-irradiation membrane damage was assessed by measuring LDH release. Cytotoxicity analysis revealed a decrease in LDH production in breast cancer cells after treatment with light doses of 5–40 J cm^{-2} at all three wavelengths. In breast CSCs, a drop in LDH production was only observed after cells were exposed to 10 J cm^{-2}. These findings revealed that light doses of 5–40 J cm^{-2} are not sufficient to induce cell death in both breast cancer cells and CSCs. On the other hand, membrane damage indicated by a significant increase in LDH production was observed in both breast cancer cells and CSCs after treatment using HF-PBM at 200 J cm^{-2} with wavelengths of 636 and 825 nm. This confirmed that the cellular response to HF-PBM differs from one cell type to another.

The concept of a tumor consisting of a population of genetically diverse cells was proposed by the CSC hypothesis in 1937 [51]. Research in the 1990s concentrated on stem cell biology, introducing the term 'cancer stem cells' for a subpopulation of cells that were found in a leukemia cell line. This subpopulation from the leukemia cell line presented with characteristics such as tumorigenicity, self-renewal, and multipotency, all properties seen in adult stem cells (ASCs) [52]. The exploitation on CSC work remains essential as it has a direct impact on current cancer therapies. The existence of CSCs cannot be proven for all cancer types, but they have been found in the following cancers: liver, brain, pancreas, colon, and ovaries [53], as well as lung cancer.

CSCs are believed to be resistant to chemotherapy and radiation therapy, this is said to be a direct result of having ASC properties. The significance of therapy-resistant CSCs leads to the exploration of therapies which can target not only cancer but also CSCs. Computer simulations have identified that non-targeting of CSCs can enrich CSC development, having a detrimental effect on the treatment outcome [54].

Present day cancer therapies are aimed at destroying the majority of the tumor, which consists of differentiated adult cancer cells capable of rapid proliferation. This in turn affects treatment outcomes as these cells capable of tumor development are not killed leading to cancer relapse and metastasis. This is due to CSCs protecting themselves in the same way normal SCs do, evidently evading treatment. Therefore, the design of a cancer treatment needs to consider the effectiveness of destroying cancerous cells as well as CSCs [55].

Lung cancer of different subtypes having the highest morbidity rate worldwide is known for its resistance against therapy and this may be related to lung CSCs residing within tumors. There have been indications that lung cancer can be propagated and maintained even by only a small population of CSCs. This statement is supported by the research findings shown by Virginia et al [56] where analysis of non-small cell lung cancer (NSCLC) containing CSCs were tested. Results indicated that CSCs make up a small proportion of the total tumor cell population. This finding that the CD133 antigen was expressed in a small percentile of the NSCLC supported the hypothesis of the existence of CSCs.

There are several biochemical mechanisms whereby PBM acts upon tissues. During tissue photobiology photons are absorbed by electronic absorption bands belonging to a chromophore or photo acceptor [57]. Tissues have been found to have optical properties called the optical window. Differing wavelengths are either absorbed or scattered by cells or tissue. Principle tissue chromophores (hemoglobin and melanin) absorb light maximally at wavelengths shorter than 600 nm while water begins to absorb significantly at wavelengths greater than 1150 nm. The optical window covers the red and near-infrared (NIR) wavelengths, where the effective tissue penetration of light is optimal. It was suggested in 1989 that the mechanism of PBM at the cellular level was based on the absorption of monochromatic, visible, and NIR radiation by components of the cellular respiratory chain [58]. The inner mitochondrial membrane contains integral membrane proteins which shuttle electrons from NADH and $FADH_2$ to oxygen molecules to form water. Energy released by this transfer to the pumping of protons (H+) from the matrix to the intermembrane space creates a gradient of protons that can flow back down this gradient, re-entering the matrix only through another complex of integral proteins in the inner membrane, the ATP synthase complex [59].

It was proposed that cytochrome c oxidase (CCO) is the primary photo acceptor for the red–NIR range in mammalian cells [60] and this was supported by results where absorption spectra obtained for CCO in different oxidation states were recorded and found to be very similar to the action spectra for biological responses to light. It absorbs light between 630 and 900 nm. A study from Pastore *et al* [61] examined the effect of He–Ne laser illumination (632.8 nm) on the purified CCO enzyme, and found increased oxidation of cytochrome c and increased electron transfer. Artyukhov and colleagues [62] found increased enzyme activity of catalase after He–Ne laser illumination. The absorption of photons by molecules leads to electronically excited states, and consequently can lead to an acceleration of electron transfer reactions [63], leading to increased production of ATP. In addition to the CCO-mediated increase in ATP production, other mechanisms may be operating in PBM. Certain molecules with visible absorption bands, such as porphyrins and some flavoproteins, can absorb photons [64, 65]. The light energy absorbed causes singlet-state excitation of oxygen molecules leading to triplet-state excitation causing an energy transfer to ground-state molecular oxygen (a triplet) to form the reactive species, singlet oxygen. Alternatively, superoxide may be formed due to electron reduction. This is the same molecule utilized in PDT to kill cancer cells. Changes in mitochondrial metabolism, and activation of the respiratory chain by irradiation using light would also increase the production of superoxide anions, O_{2-}. All of the responses were involved in the cell's redox states. Several regulation pathways are mediated through the cellular redox state which induce the activation of numerous intracellular signaling pathways, regulate nucleic acid synthesis, protein synthesis, enzyme activation, and cell-cycle progression [66].

References

[1] Yu C H and Yu C C 2014 Photodynamic therapy with 5-aminolevulinic acid (ALA) impairs tumor initiating and chemo-resistance property in head and neck cancer-derived cancer stem cells *PloS one* **9** e87129

[2] Wei M F *et al* 2014 Cell death of colorectal cancer stem-like cell was induced by photodynamic therapy with protoporphyrin IX *Institute of Biomedical Engineering, National Taiwan University Report*

[3] de Paula L B *et al* 2015 Combination of hyperthermia and photodynamic therapy on mesenchymal stem cell line treated with chloroaluminum phthalocyanine magnetic-nano-emulsion *J. Magn. Magn. Mater.* **380** 372–6

[4] Gottesman M M and Ambudkar S V 2001 Overview: ABC transporters and human disease *J. Bioenerg. Biomembr.* **33** 453–8

[5] Vasiliou V, Vasiliou K and Nebert D W 2009 Human ATP-binding cassette (ABC) transporter family *Human Genom* **3** 281–90

[6] Bailey-Dell K J *et al* 2001 Promoter characterization and genomic organization of the human breast cancer resistance protein (ATP-binding cassette transporter G2) gene *Biochim. Biophys. Acta.* **1520** 234–41

[7] Halwachs S *et al* 2016 The ABCG2 efflux transporter from rabbit placenta: Cloning and functional characterization *Placenta* **38** 8–15

[8] Filia M F *et al* 2017 Induction of ABCG2/BCRP restricts the distribution of zidovudine to the fetal brain in rats *Toxicol. Appl. Pharmacol.* **330** 74–83

[9] Vlaming M L, Lagas J S and Schinkel A H 2009 Physiological and pharmacological roles of ABCG2 (BCRP): recent findings in Abcg2 knockout mice *Adv. Drug Deliv. Rev.* **61** 14–25

[10] Tian Y *et al* 2015 Interplay of breast cancer resistance protein (BCRP) and metabolizing enzymes *Curr. Drug Metab.* **16** 877–93

[11] van Herwaarden A E *et al* 2007 Multidrug transporter ABCG2/breast cancer resistance protein secretes riboflavin (vitamin B2) into milk *Mol. Cell. Biol.* **27** 1247–53

[12] Ding X W, Wu J H and Jiang C P 2010 ABCG2: a potential marker of stem cells and novel target in stem cell and cancer therapy *Life Sci.* **86** 631–7

[13] Robey R W *et al* 2007 ABCG2: determining its relevance in clinical drug resistance *Cancer Metastasis Rev.* **26** 39–57

[14] Seigel G M *et al* 2005 Cancer stem cell characteristics in retinoblastoma *Mol. Vis.* **11** 729–37

[15] Ho M M *et al* 2007 Side population in human lung cancer cell lines and tumors is enriched with stem-like cancer cells *Cancer Res.* **67** 4827–33

[16] Shi G M *et al* 2008 Identification of side population cells in human hepatocellular carcinoma cell lines with stepwise metastatic potentials *J. Cancer Res. Clin. Oncol.* **134** 1155–63

[17] Wang Y H *et al* 2009 A side population of cells from a human pancreatic carcinoma cell line harbors cancer stem cell characteristics *Neoplasma* **56** 371–8

[18] Monzani E *et al* 2007 Melanoma contains CD133 and ABCG2 positive cells with enhanced tumourigenic potential *Eur. J. Cancer* **43** 935–46

[19] Olempska M *et al* 2007 Detection of tumor stem cell markers in pancreatic carcinoma cell lines *Hepatobiliary Pancreat. Dis. Int.* **6** 92–7

[20] Zen Y *et al* 2007 Histological and culture studies with respect to ABCG2 expression support the existence of a cancer cell hierarchy in human hepatocellular carcinoma *Am. J. Pathol.* **170** 1750–62

[21] Cowled P A and Forbes I J 1989 Modification by vasoactive drugs of tumour destruction by photodynamic therapy with haematoporphyrin derivative *Br. J. Cancer* **59** 904–9

[22] Gossner L *et al* 1991 Verapamil and hematoporphyrin derivative for tumour destruction by photodynamic therapy *Br. J. Cancer* **64** 84–6

[23] Jonker J W *et al* 002 The breast cancer resistance protein protects against a major chlorophyll-derived dietary phototoxin and protoporphyria *Proc. Natl. Acad. Sci. USA* **99** 15649–54

[24] Busch T M and Hahn S M 2005 Multidrug resistance in photodynamic therapy *Cancer Biol. Ther.* **4** 195–6

[25] Robey R W *et al* 2005 ABCG2-mediated transport of photosensitizers: potential impact on photodynamic therapy *Cancer Biol. Ther.* **4** 187–94

[26] Liu W *et al* 2007 The tyrosine kinase inhibitor imatinib mesylate enhances the efficacy of photodynamic therapy by inhibiting ABCG2 *Clin. Cancer Res.* **13** 2463–70

[27] Morgan J *et al* 2010 Substrate affinity of photosensitizers derived from chlorophyll-a: the ABCG2 transporter affects the phototoxic response of side population stem cell-like cancer cells to photodynamic therapy *Mol. Pharm.* **7** 1789–804

[28] Pan L *et al* 2017 The sensitivity of glioma cells to pyropheophorbide-alphamethyl ester-mediated photodynamic therapy is enhanced by inhibiting ABCG2 *Lasers Surg. Med.* **49** 719–26

[29] Salama R, Tang J, Gadgeel S M, Ahmad A and Sarkar F H 2012 Lung cancer stem cells: current progress and future perspectives *J. Stem Cell Res. Ther.* **S7** 007

[30] Tirino V, Desideri V, Paino F, Papaccio G and De Rosa M 2012 Methods for cancer stem cell detection and isolation *Methods Mol. Biol.* **879** 513–29

[31] Ren F, Sheng W Q and Du X 2013 CD133: a cancer stem cells marker, is used in colorectal cancers *World J. Gastroenterol.* **19** 2603–11

[32] Crous A 2016 Photobiomodulatory effects of low intensity laser irradiation on isolated lung cancer stem cells *Master's thesis* Available at: http://hdl.handle.net/10210/226444

[33] Mvula B, Mathope T, Moore T and Abrahamse H 2008 The effect of low level laser irradiation on adult human adipose derived stem cells *Lasers Med. Sci.* **23** 277–82

[34] De Villiers J, Houreld N and Abrahamse H 2011 Influence of low intensity laser irradiation on isolated human adipose derived stem cells over 72 hrs and their differentiation potential into smooth muscle cells using retinoic acid *Stem Cell Rev. Rep.* **7** 869–82

[35] Caligur V 2008 The cancer stem cell hypothesis *Sigma Aldrich report* available at www.sigmaaldrich.com/technical-documents/articles/biofiles/the-cancer-stem-cell.html

[36] Abrahamse H 2010 Low intensity laser irradiation ameliorates stem cell based therapy for use in autologous grafts *ScienceMED* **1** 1–6

[37] Croker A K and Allan A L 2008 Cancer stem cells: implications for the progression and treatment of metastatic disease *J. Cell. Mol. Med.* **12** 374–90

[38] Moore P, Ridgway T D, Higbee R G, Howard E W and Lucroy M D 2005 Effect of wavelength on low-intensity laser irradiation stimulated cell proliferation *in vitro Lasers Surg. Med.* **36** 8–12

[39] Hu W P, Wang J J, Yu C L, Lan C C, Chen G S and Yu H S 2007 Helium–neon laser irradiation stimulates cell proliferation through photostimulatory effects in mitochondria *J. Investig. Dermatol.* **127** 2048–57

[40] Crous A and Abrahamse H 2016 High fluence low intensity laser irradiation bioinhibits viability and proliferation of lung cancer stem cells *Stem Cell Res. Ther.* **6** 368

[67] Crous A and Abrahamse H 2016 Low intensity laser irradiation at 636nm induces increased viability and proliferation in isolated lung cancer stem cells *Photomed. Laser Surg.* **34** 525–32

[41] Lavi R, Shainberg A, Friedmann H, Shneyvays V, Rickover O, Eichler M, Kaplan D and Lubart R 2003 Low energy visible light induces reactive oxygen species generation and

stimulates an increase of intracellular calcium concentration in cardiac cells *J. Biol. Chem.* **278** 40917–22

[42] Ocana-Quero J M, Perez de la Lastra J, Gomez-Villamandos R and Moreno-Millan M 1998 Biological effect of helium–neon (He–Ne) laser irradiation on mouse myeloma (Sp2-Ag14) cell line *in vitro Lasers Med. Sci.* **13** 214–8

[43] Rahman M, Deleyrolle L, Vedam-Mai V, Azari H, Abd-El-Barr M and Reynolds B A 2011 The cancer stem cell hypothesis: failures and pitfalls *Neurosurgery* **68** 531–45

[44] Manoto S L, Sekhejane P R, Houreld N N and Abrahamse H 2012 Localization and phototoxic effect of zinc sulfophthalocyanine photosensitizer in human colon (DLD-1) and lung (A549) carcinoma cells (*in vitro*) *Photodiag. Photodyn. Ther.* **9** 52–9

[45] Chen Y C, Hsu H S, Chen Y W, Tsai T H, How C K and Wang C Y 2008 Oct-4 expression maintained cancer stem-like properties in lung cancer-derived CD133-positive cells *PLoS One* **3** e2637

[46] Takenobu H, Shimozato O, Nakamura T, Ochiai H, Yamaguchi Y, Ohira M, Nakagawara A and Kamijo T 2011 CD133 suppresses neuroblastoma cell differentiation via signal pathway modification *Oncogene* **30** 97–105

[47] Ferreira C G, Epping M, Kruyt F A and Giaccone G 2002 Apoptosis: target of cancer therapy *Clin. Cancer Res.* **8** 2024–34

[48] Sun H, Jia J, Wang X, Ma B, Di L, Song G and Ren J 2013 CD44+/CD24- breast cancer cells isolated from MCF-7 cultures exhibit enhanced angiogenic properties *Clin. Transl. Oncol.* **15** 46–54

[49] Pannuti A *et al* 2010 Targeting cancer stem cells through notch signaling *Clin. Cancer Res.* **16** 3141–52

[50] Al-Watban F A and Andres B L 2012 Laser biomodulation of normal and neoplastic cells *Lasers Med. Sci.* **27** 1039–43

[51] Clevers H 2011 The cancer stem cell: premises, promises, and challenges *Nat. Med.* **17** 313–9

[52] Cherryholmes G 2013 Cancer stem cell hypothesis: proceed with caution *Saguine Biosciences blog* available at http://technical.sanguinebio.com/cancer-stem-cell-hypothesis-proceed-with-caution/

[53] Magee J A, Piskounova E and Morrison S J 2012 Cancer stem cells: impact, heterogeneity, and uncertainty *Cancer Cell* **21** 283–96

[54] Vermeulen L, de Sousa e Melo F, Richel D J and Medema J P 2012 The developing cancer stem-cell model: clinical challenges and opportunities *Lancet Oncol.* **13** 83–9

[55] University of Pittsburgh Schools of the Health Sciences 2007 Cancer stem cells similar to normal stem cells can thwart anti-cancer agents *Science Daily* Available at www.sciencedaily.com /releases/2007/06/070615110244

[56] Virginia T, Rosa C, Renato F, Donatella M, La Roccae A, Giuseppe V, Gaetano R and Giuseppe P 2009 The role of CD133 in the identification and characterisation of tumour-initiating cells in non-small-cell lung cancer *Eur. J. Cardio-Thorac. Surg.* **36** 446–53

[57] Sutherland J C 2002 Biological effects of polychromatic light *Photochem. Photobiol.* **76** 164–70

[58] Karu T I 1989 Photobiology of low-power laser effects *Health Phys.* **56** 691–704

[59] Hamblin M R 2008 Mechanisms of low level light therapy *Photobiological Sciences Online* available at www.photobiology.info/Hamblin.html

[60] Karu T I and Afanaseva N I 1995 Cytochrome c oxidase as the primary photoacceptor upon laser exposure of cultured cells to visible and near IR-range light *Dokl. Akad. Nauk.* **342** 693–5

[61] Pastore D, Greco M and Passarella S 2000 Specific helium–neon laser sensitivity of the purified cytochrome c oxidase *Int. J. Radiat. Biol.* **76** 863–70

[62] Artyukhov V G, Basharina O V, Pantak A A and Sveklo L S 2000 Effect of helium–neon laser on activity and optical properties of catalase *Bull. Exp. Biol. Med.* **129** 537–40

[63] Yu W, Naim J O, McGowan M, Ippolito K and Lanzafame R J 1997 Photomodulation of oxidative metabolism and electron chain enzymes in rat liver mitochondria *Photochem. Photobiol.* **66** 866–71

[64] Friedmann H, Lubart R, Laulicht I and Rochkind S 1991 A possible explanation of laser-induced stimulation and damage of cell cultures *J. Photochem. Photobiol. B.* **11** 87–91

[65] Eichler M, Lavi R, Shainberg A and Lubart R 2005 Flavins are source of visible-light-induced free radical formation in cells *Lasers Surg. Med.* **37** 314–9

[66] Liu H, Colavitti R, Rovira I I and Finkel T 2005 Redox-dependent transcriptional regulation *Circulat. Res.* **97** 967–74

IOP Concise Physics

Photomedicine and Stem Cells
The Janus face of photodynamic therapy (PDT) to kill cancer stem cells, and photobiomodulation
(PBM) to stimulate normal stem cells
Heidi Abrahamse and Michael R Hamblin

Chapter 4

Normal stem cells

4.1 Developmental biology

Our knowledge about the physiology and anatomy of the human genome is substantial, while significant information about the regulation of embryonic development is still required. Model organisms have been used to study human development [1] although stem cells are the foundation of all tissues in your body. Different types of stem cells are found and differentiate at different times throughout the body during our life-span. Embryonic stem cells (ESCs) exist only at the earliest stages of development, while tissue-specific adult stem cells (ASCs) form during foetal development and remain in our bodies throughout life. All stem cells can self-renew and differentiate [2].

Stem cells have the capacity for long-term self-renewal without senescence and the ability to differentiate into one or more specialized cell types, thus providing an inexhaustible supply of cells for transplantation (pluripotency). Totipotent stem cells can generate all tissue types and play an important role in human development. They provide the raw material for the development of all tissues and organs in the embryo and all the extra-embryonic tissues. Tissue-specific stem cells are deposited in niches throughout the body, such as bone marrow (BM), brain, liver, and skin, as a mechanism for tissue maintenance, growth, and repair in later life [3–6]. ASCs thought to regenerate only a restricted set of cell lineages have now been shown to have more plasticity [7–9].

Embryonic carcinoma (EC) cell lines, derived from the undifferentiated compartment of murine and human germ cell tumors, were the first to be cultured as pluripotent cell lines [10, 11]. Human ESC lines have caused strong ethical and political controversy since human embryos and their use for scientific research remains objectionable to many religious communities [12]. Legislation governing the use of human embryos to produce ESC lines varies between countries. Somatic nuclear transfer, or cloning, the fusion of sperm and egg is an alternative method of producing human ESCs. Autologous cloned human embryos can prevent potential rejection and hold significant promise for transplantation medicine [13].

An ASC is undifferentiated, can renew itself, and can differentiate into the major specialized cell types of the tissue or organ. The primary roles of ASCs are to maintain and repair the tissue in which they are found. Also termed somatic stem cells, ASCs refer to cells of the body (not the germ cells, sperm, or eggs) and unlike ESCs, the origin of ASCs is mature tissues. ASCs are found in many more tissues than once thought possible, thus generating excitement for the idea of ASCs being used for transplants. Adult hematopoietic stem cells from BM have been used in transplants for more than 40 years. Evidence exist of stem cells in the brain and the heart, where ASCs were not at first expected. If the differentiation of ASCs can be induced and controlled in the laboratory, they may become the basis of transplantation-based therapies. In the 1950s, researchers discovered that BM contains two kinds of stem cells called hematopoietic stem cells which can form all the types of blood cells in the body and BM stromal stem cells (also called MSCs). Non-hematopoietic stem cells can generate bone, cartilage, and fat cells that support the formation of blood and fibrous connective tissue. In the 1960s, scientists discovered two regions of the brain in rats that contained dividing cells that become nerve cells. Most scientists believed that the adult brain could not generate new nerve cells, but in the 1990s scientists established that the adult brain contains stem cells that are able to generate the brain's three major cell types—astrocytes and oligodendrocytes, which are non-neuronal cells, and neurons, or nerve cells [14].

The first human ESC lines from cultured human blastocysts were formed by Thomson and colleagues in 1998 [15], and Yamanaka and Takahashi reprogrammed adult mouse cells into pluripotent stem cells by expressing four transcription factors [Oct3/4 (Pou5f1), Sox2, Myc and Klf4] [16]. The step to human pluripotent stem cells (hPSCs) took only one year, and human-induced pluripotent stem cells (hiPSCs) were soon generated [17, 18]. This made it possible to study crucial aspects of a disease and use for autologous cell replacement therapy [19]. *In vitro* differentiation and genetic manipulation of hPSCs also provide great opportunities to study human embryonic development since hPSCs and hiPSCs can self-renew indefinitely while maintaining their differential capacity to become almost any cell type in the human body [15, 17, 18].

hPSCs are well suited for studies of human development since they have the potential to generate every adult cell type. At the same time, the *in vitro* culture system allows a rapid and cost-effective model to investigate the genetic function of a developmental process. Additionally, the unlimited self-renewal capacity of hPSCs provides abundant material for screening and can thus be used to test outcomes or postulations of previous studies in organisms, as well as new research through biological and chemical screening, but still requiring *in vitro* differentiation platforms that mimic embryonic development [1].

The maintenance of pluripotency in hPSCs has improved our understanding of the mechanisms of signal transduction, and the role that transcription factors and epigenetic regulators play during development in general. Studies using hESCs, zebrafish, and flies yielded the process of transcriptional pausing after promoter binding and transcription initiation through genome-wide analysis of histone modifications [20]. In addition, it indicated that transcriptional pausing is also present in differentiated cell types and may contribute to cell fate determination during embryonic development [21, 22]. hPSCs have also been used as a model system for investigating the mechanisms that underlie embryonic development. Genetic screens and loss- and gain-of-function (LOF and GOF) studies, in the mouse and other model organisms, have identified many genes and signaling pathways that govern various aspects of development. Investigation for defined conditions to differentiate hPSCs into specific cell types of all three germ layers (Ec, ectoderm; Me, mesoderm; En, endoderm) was done as a direct result of identifying mechanisms of embryonic development using hPSCs. The generation of specific cell types from hPSCs has advanced regenerative medicine including cell replacement therapy, disease modelling, and drug discovery (figure 4.1).

To evaluate embryonic development hPSCs can be used to evaluate the process *in vitro* although human specific transcriptional regulation and signaling pathways have been identified that may contribute to this process. For example, neuro-ectoderm specification is a process of gastrulation where Pax6 is required for neuroectoderm specification from hESCs but not from mESCs [23]. hPSCs can also be used for studying conserved developmental mechanisms—e.g. the exact role of transforming growth factor-β (TGF-β) signaling in pancreatic development has been questioned in mice [24]—owing to the time-sensitive requirement of TGF-β signaling during pancreatic development. TGF-β signaling is used repeatedly during

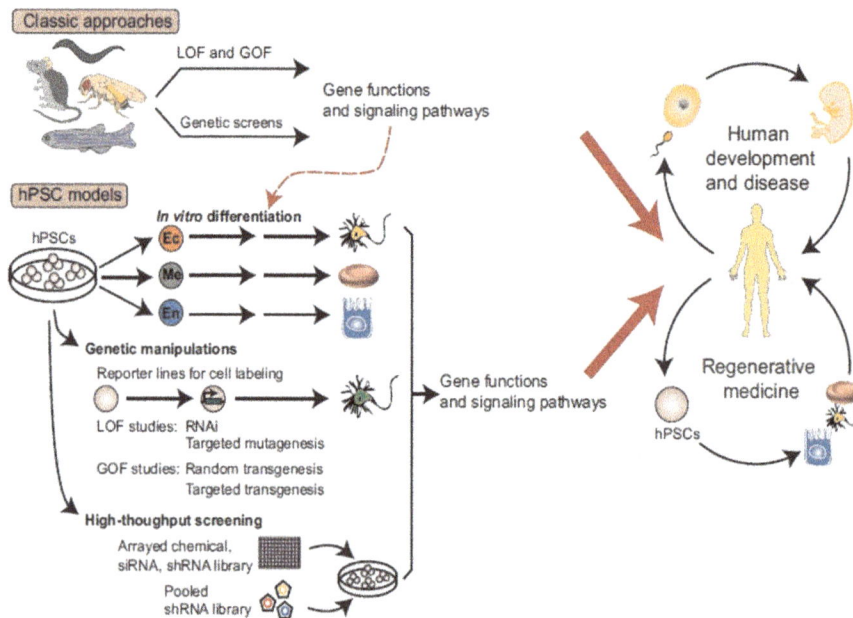

Figure 4.1. Advancing developmental biology and regenerative medicine through studies of hPSCs and model organisms.Reproduced from [1]. Copyright 2013 with permission of The Company of Biologists.

embryonic development. To avoid the limitations of pleiotropic gene functions in *in vivo* studies, the generation of conditional knockout or hypomorphic alleles makes it easier to manipulate TGF-β signaling using hPSCs. hPSC-based studies showed that TGF-β signaling inhibits the differentiation of pancreatic progenitors into the endocrine lineage—the lineage that gives rise to β-cells [25, 26]. These studies provide evidence of the useful model of hPSCs to determine both conserved and non-conserved developmental mechanisms [1].

4.2 Markers and phenotypes

ESC markers are molecules specifically expressed in ESCs and used to characterize and study the mechanism of ESC pluripotency and their maintenance and self-renewal. Different cell types sometimes share single or multiple markers, making this the main obstacle in the clinical application of ESC purification. Marker-based flow cytometry and magnetic cell sorting are the most effective cell isolating methods. There is a wide range of cell surface and generic molecular markers as well as other molecules, such as lectins and peptides, which bind to ESCs via affinity and specificity that are able to identify undifferentiated ESCs [27].

Proteins involved in several signal pathways are also known to have important functions in cell fate decision while lectins and other peptides specifically bind to ESCs. However, many ESC markers overlap with tumor stem cell markers. Understanding the mechanisms that regulate the pluripotency of hESCs is a challenge, since research has shown that human and mouse ESCs differ in these mechanisms despite their similar embryonic origins [28]. Insight into marker specificity is critical for proper use and elucidating the mechanisms regulating pluripotency and self-renewal of ESCs [27], see figure 4.2.

Stem cell therapy has become a promising tool for studying the mechanisms of development, regeneration, and treatment modality for a variety of disorders. Stem cells are found in the embryo and most adult tissues that are active in endogenous tissue regeneration. They are capable of autorenovation, often maintain their multipotency of differentiation into various tissues of their germ line, and are perfect candidates for cellular therapy due to the fact that they can be identified and

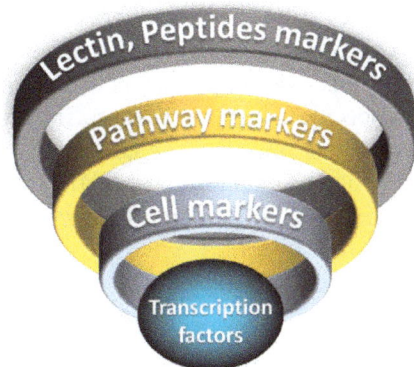

Figure 4.2. Categories of ESC markers [27].

isolated. Stem cell marker expression is used for identification of various stem cells including mesenchymal, epithelial, and limbal epithelial stem cells, endothelial progenitor cells, supra adventitial adipose stromal cells, adipose pericytes, as well as cancer stem cells. A single stem cell marker cannot be used to isolate and distinguish a stem cell type because their expression partially overlaps between lineages. Flow cytometry allows the simultaneous detection of various markers which facilitate stem cell identification for clinical diagnosis and research. Several studies indicate that expression of classical markers for stem cell classification, such as CD34, CD45, and CD133, may differ between the virtually identical stem and progenitor cells, i.e. endothelial progenitor or mesenchymal stem cells, when they were obtained from different tissues. This may be due to phenotypic differences due to the source or it may be caused by different isolation and experimental conditions [29].

4.2.1 Cell surface markers

The cell surface is coated with specialized proteins that can selectively bind or adhere to other signal molecules. Different types of membrane proteins differ in structure and in their affinity for signal molecules, and some proteins are specifically found and secreted in specific cell types and are thus able to act as cell markers. Membrane proteins are the most important marker type in recognizing ESCs without breaking the cell membrane (table 4.1). However, most of the membrane markers overlap with tumor cell types.

4.2.2 Transcription factors

Transcription factors are crucial for gene regulation and therefore mostly found in an inactive form under normal conditions, binding only to their cognate recognition sequences when a specific signal transduction process occurs. Genetic expression and subsequent protein function in the nucleus indicates that the cell has responded to a certain condition. Tracking gene expression can be used as a marker for a specific cell situation. Transcription factors that expressed in ESC are listed in table 4.2.

4.2.3 Signal pathway-related intracellular markers

The most significant markers exist on the cell surface in order for an external signal to penetrate through the cellular membrane and transmit signals. When a specific intracellular signal pathway is activated, it triggers a series of molecular events that leads the regulation of specific gene expression in response to this signal. Intracellular signal pathways play important roles in maintaining ESC self-renewal and pluripotency; therefore, many signal pathway-related proteins, which are critical for these roles, can be regarded as ESC markers. The potential markers among these pathways are shown in table 4.3.

4.2.4 Enzymatic markers

Alkaline phosphatase staining is used to detect both mouse and human ESCs to test pluripotency since both cell types express high levels of alkaline phosphatase and

Table 4.1. ESC surface markers [27].

	Characteristics	Classification	References
SSEAs markers			
SSEA-1 (CD15Lewis x)	Murine embryos, mouse ES cells, mouse and human germ cells, EC cells	Carbohydrate-associated molecules	[30–37]
SSEA-3	Primate ES cells, human embryonic germ cells, human ES cells, EC cells	Carbohydrate-associated molecules	
SSEA-4	Primate ES cells, human embryonic germ cells, human ES cells, EC cells	Carbohydrate-associated molecules	
CD markers			
CD324 (E-Cadherin)	Human ES cells, mouse ES cells, EC cells	Surface marker (binding to integrin alphaE/beta7, homotypic interactions mediate cell adhesion)	[38–47]
CD90 (Thy-1)	Human ES cells, mouse ES cells, hematopoietic stem cells, EC cells	Surface marker (hematopoietic stem cell and neuron differentiation, T activation)	
CD117 (c-Kit, SCFR)	Human ES cells, mouse ES cells, hermatopoietic stem progenitors, neural crest-derived melanocytes, primordial germ cells, EC cells	Surface maker (Stem cell factor receptor)	
CD326	Human ES cells, mouse ES cells, EC cells	Surface maker (function as growth factor receptor or adhesion molecule)	
CD9 (MRP1, TM4SF DRAP-27, p24)	Human ES cells, mouse ES cells	Surface maker (cell adhesion, migration T co-stimulation)	
CD29 (β1 integrin)	Human ES cells, mouse ES cells	Surface maker	
CD25 (HAS)	Human ES cells, mouse ES cells	Surface maker (T co-stimulation, CD62P receptor)	
CD59 (Protectin)	Human ES cells, mouse ES cells	Surface maker (binds complement C8 and C9, blocks membrane attack complex assembly)	

CD133	Human ES cells, mouse ES cells, EC cells, hematopoietic stem cells	Surface maker	
CD31 (PECAM-1)	Human ES cells, mouse ES cells	Surface maker (CD38 receptor, signaling, platelet-endoth adhesion)	
CD49f (Integrin α6/CD29)	Human ES cells, mouse ES cells	Membrane receptors	[48–57]
Markers			
TRA-1-60	Human ES cells, teratocarcinoma, embryonic germ cells, EC cells	Surface antigen	[34, 41, 58–60]
TRA-1-81	Human ES cells, teratocarcinoma, embryonic germ cells, EC cells	Surface antigen	
Frizzled5	Human ES cells, mouse ES cells	Seven transmembrane-spanning G-protein-coupled receptor	[61–65]
Stem cell factor (SCF of c-Kit ligand)	ES cells, mouse ES cells, hematopoietic stem cells, mesenchymal stem cells, EC cells	Cytokine, exist both as a transmembrane protein and a soluble protein	[66, 67]
Cripto (TDGF-1)	Mouse ES cells, human ES cells, cardiomycocyte, EC cells	Receptor for the TGF-β signaling pathway	[68, 69]

Table 4.2. Transcription factors [27].

CORE nuclear transcription factor	Characteristics	Classification	References
Oct-3/4 (Pou5fl)	Mouse ES cells, human ES cells, EC cells	POU family, transcription factors	[70, 71]
Sox2	Mouse ES cells, human ES cells, EC cells, NS cells	POU family binder, transcription factors	[72, 73]
KLF4	Mouse ES cells, human ES cells, EC cells	Zinc-finger, transcription factors	[74]
Nanog	Mouse ES cells, human ES cells, EC cells	Transcription factor	[75–77]
Markers			
Rex1 (Zfp42)	Mouse ES cells, human ES cells, EC cells	Zinc-finger, transcription factor	[78–80]
UTF1	Mouse and human ES cells, germ line tissues in mouse and human, EC cells	Transcriptional coactivator	[81, 82]
ZFX	Murine ES cells, human ES cells, hematopoietic stem cells, EC cells	X-linked zinc-finger protein, probable transcriptional activators	[83, 84]
TBN	Mouse, human inner cell mass	New class of proteins with an important function in development	[85]
FoxD3	Murine ES cells, human ES cells, EC cells	Forkhead Box family, transcriptional regulator	[86–88]
HMGA2	Mouse ES cells, human ES cells	Architectural transcription factor	[89–92]
NAC1	Mouse ES cells, human ES cells	The POZ/BTB domain family, nuclear factor	[93–95]
GCNF (NR6A1)	Mouse ES cells, human ES cells, EC cells	Nuclear receptor gene superfamily, nuclear receptor	[96, 97]
Stat3	Murine ES cells, human ES cells, EC cells	Transcription factor	[98, 99]
LEF1, TCF3	Mouse ES cells, human ES cells, EC cells	(HMG) DNA-binding protein family, transcription factor	[100]
Sall4	Murine ES cells, Human ES cells, EC cells	Zinc-finger, transcription factor	[101, 102]
Fbxo15	Mouse ES cells, early embryos and testis tissue, EC cells	F-box protein family, target of Oct3/4	[103]
ECAT genes			
ECAT11 (FLJ10884/ L1TD1)	Human ES cells, EC cells	Downstream target of Nanog	[104]

Ecat1	Mouse oocytes, EC cells	KH domain containing RNA-binding protein	[105]
ECAT9 (Gdf3)	Human ES cells, EC cells	TGF-β superfamily, BMP inhibitor	[106]
Dppa genes	**Oct4-related genes**		
Dppa5 (ESG1)	Mouse ES cells, human ES cells, EC cells	K-homology RNA-binding (KH) domain	[107, 108]
Dppa4	Mouse ES cells, human ES cells, EC cells	Nuclear factor	[109]
Dppa2 (ECSA)	Mouse ES cells, human ES cells, EC cells	DNA-binding protein	[110, 111]
Dppa3 (Stella)	Mouse ES cells, human ES cells, EC cells, primordial germ cells, oocytes, preimplantation embryos	Maternal factor	[112, 113]

Table 4.3. Signal pathway-related intracellular markers [27].

Markers	Characteristics	Classification	References
SMAD2/3	Human ES cells, EC cells	Smad proteins (R-Smad, TGF-β/activin/nodal signaling pathway)	
β-catenin	Mouse ES cells, human ES cells, EC cells	Transcription activators, Wnt/β-catenin signaling pathway	[114, 115]
SMAD1/5/8	Mouse ES cells, EC cells	Smad proteins (R-Smad), BMP signaling pathway	[116–118]
SMAD4	Mouse ES cells, human ES cells, EC cells, early embryos and testis tissue	Smad proteins (Co-SMAD), TGF-β/activin/nodal signaling pathway, BMP signaling pathway	

telomerase. ESCs have elevated levels of alkaline phosphatase on their cell membrane; in humans TRA-2-49 and TRA-2-54 antibodies can detect the alkaline phosphatase. In murine cells, they are visualized by an enzymatic-based reaction [61]. These markers are listed on NIH stem cell resource web site[1].

Stem cell lineages are identified by their typical tissue-specific localization, expression of specific stemness marker proteins and other stem cell specific epitopes, which are not expressed by somatic cells. Because of overlapping expression patterns between stem cell lineages, a stem cell can often not be classified by detection of a single marker protein. Multi-parameter analysis of six to eight markers at the same time is now possible due to monoclonal antibodies coupled to fluorochromes with excitation wavelengths proximal to multiple laser lines of flow cytometers [119]. Subsequently, multi-color detection by flow cytometry has been turned into an ideal tool for the identification and purification of stem cells. Analysis of a combination of lineage-positive and -negative marker proteins allows one to distinguish between hematopoietic and endothelial stem cells by simultaneous polychromatic detection of the expression of various antigens [29].

4.3 Stem cell niches

Stem cells participate in physiologic systems that determine the outcome of developmental events and stress. These cells are fundamental to tissue maintenance and repair and the signals they receive play a critical role in the integrity of the organism. Although substantial resources have been focused on stem cell identification and the molecular pathways involved in their regulation, how these pathways achieve physiologically responsive stem cell functions is still unclear. Our understanding of stem cells in the context of their microenvironment and the relation between stem cell niche dysfunction, carcinogenesis, and aging still requires extensive attention [120].

---

[1] http://stemcells.nih.gov/info/scireport/appendixe.asp#eii.

Stem cell niches are formed during ontogeny and can remain unoccupied and exist independently of stem cells; however, stem cell self-renewal cannot be maintained for long periods outside of the niche except for particular conditions. The nature of the stem cell niche and its interaction with stem cells is a fundamental question that needs to be addressed by research. A vacant niche can be occupied by excessive or transplanted stem cells and can provide for their functioning. A niche size allows a definite number of stem cells to be maintained. Excessive stem cells either differentiate in the presence of specific signal(s) or undergo apoptosis in the absence of such a signal. The niches control the number of stem cells and protect it from excessive stem cell proliferation although under certain conditions, stem cells can leave and return to their niches. Stem cells are retained in the niche by cell-to-cell interactions and adhesion to the extracellular matrix (ECM). Both the niches and stem cells arise at a particular ontogenetic stage and are capable of long self-renewal. The development can be described in terms of the formation of stem cells and their niches [121].

The stem cell niche is the *in vivo* microenvironment and should not be considered simply a physical location for stem cells, but as the place where extrinsic signals interact and integrate to influence stem cell behavior. The stimuli include cell-to-cell and cell matrix interactions and signals (molecules) that activate and/or repress genes and transcription programs. A direct consequence of this interaction is that stem cells are maintained in a dormant state, induced to self-renewal, or move to a differentiated state. Schofield first postulated the hypothesis of a specialized stem cell microenvironment in 1978 [122]. He speculated that niches have an anatomical location and removal of stem cells from their niche results in differentiation. The first demonstration and characterization of niche components was conducted in the invertebrate models of *Drosophila melanogaster* and *Caenorhabditis elegans* gonads [123, 124]. Examination of these systems showed that the fundamental anatomical components and molecular pathways of the niche environment are highly conserved among species, although their respective roles within the niche may show distinct variations. Common niche components that are associated with similar functions exist and the niche model postulates the association between resident stem cells and heterologous cell types—the niche cells. However, the existence of heterologous cell types is not essential and components of the ECM can determine the niche for stem cells so that a niche environment may retain its key functions and properties, even in the temporary absence of stem cells, allowing recruitment and homing of exogenous stem cells to the pre-existing stem cell niche. Components of the niche include: stromal support cells that contain cell–cell adhesion molecules and secreted soluble factors found in close proximity to stem cells; ECM proteins that act as stem cell anchors and form a mechanical scaffold to transmit stem cell signaling; blood vessels that carry nutritional support and systemic signals to the niche and recruit stem cells from and to the niche; and finally neural inputs that favor the mobilization of stem cells out of their niches and integrate signals from different organ systems. Neuronal stimulation is important in hematopoietic stem cell trafficking [125]. Given the profound effect of the niche environment on stem cell behavior, recent studies are investigating niche distresses that cause stem cell dysfunctions such as aging or neoplastic transformation [126, 127].

The stem cell niche concept has gained experimental support and conceptual complexity since it was proposed by Schofield. Interaction between different cell types intrinsic to the stem cell niche offers the opportunity to target these cell communication networks and tailor the dynamics of normal stem cells to boost their ability to respond to injury, as well as to manage the competitive advantage of malignant cells. The niche may become a target for intervention and an opportunity to affect regenerative medicine and anticancer treatments [120].

4.4 Asymmetric and symmetric stem cell division

Stem cells give rise to daughter cells that are committed to lineage-specific differentiation by undergoing an intrinsically asymmetric cell division whereby they segregate cell fate determinants into only one of the two daughter cells. Alternatively, they can orient their division plane so that only one of the two daughter cells maintains contact with the niche and stem cell identity [128]. Stem cells are defined by both their ability to make more stem cells and their ability to produce cells that differentiate. One strategy by which stem cells can accomplish these two tasks is asymmetric cell division, whereby each stem cell divides to generate one daughter with a stem cell fate (self-renewal) and one daughter that differentiates [129–132]. Asymmetric division manages both tasks with a single division; however, a disadvantage of this type of division is that it prevents stem cell proliferation. This lack of flexibility is a problem, given that stem cell numbers can increase markedly, both when stem cell pools are first established during development [124, 133, 134] and when they are regenerated after injury [135–138]. To overcome this reduction in quantity, stem cells have additional self-renewal strategies that allow control of their numbers, known as symmetric division, that lead to self-renewal and generate differentiated progeny. Symmetric divisions are defined as the generation of daughter cells that are destined to acquire the same fate. Although the idea that stem cells can divide symmetrically may seem counterintuitive, stem cells are defined by their 'potential' to generate more stem cells and differentiated daughters, rather than by their production of a stem cell and a differentiated daughter at each division. A pool of stem cells with equivalent developmental potential may produce only stem cell daughters in some divisions and only differentiated daughters in others. Stem cells can thus divide either completely by symmetric divisions or by a combination of symmetric and asymmetric divisions (figure 4.3).

Two types of mechanisms govern asymmetric cell divisions, the first relies on the asymmetric partitioning of cell components that determine cell fate (intrinsic) and secondly asymmetric placement of daughter cells relative to external cues (extrinsic). Intrinsic mechanisms involve regulated assembly of cell polarity factors and regulated segregation of cell fate determinants. Daughter cells may have equivalent developmental potential, but acquire different fates due to exposure to varying external signals leading to asymmetric division. The division is asymmetric with respect to the ultimate fate of the daughter cells although the division is intrinsically symmetric, initially yielding two daughter cells with equivalent developmental

Asymmetric Division

Stem

Diff   Stem

Symmetric Division

Stem

Stem   Stem

Stem

Diff   Diff

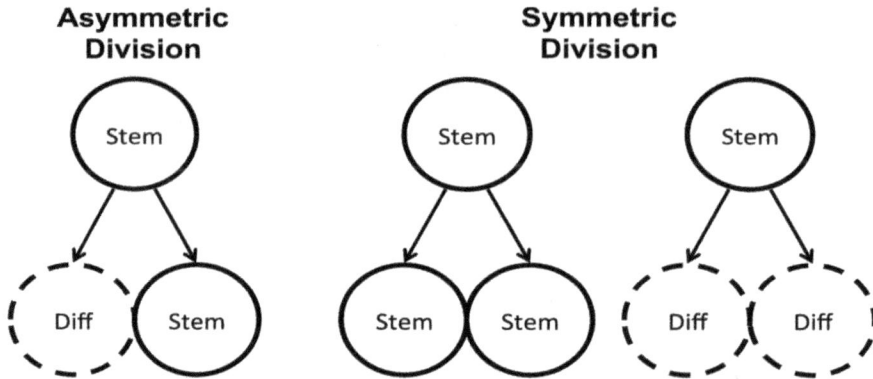

Figure 4.3. Symmetric and asymmetric stem cell divisions [139].

potential. For many asymmetric divisions, the mitotic spindle is regulated so that its orientation is reproducible—a process that can be controlled by both extrinsic and intrinsic cues. An example of asymmetric division that is controlled by an intrinsic mechanism is found in *C. elegans* zygotes, that divides asymmetrically to produce one larger blastomere to make ectoderm, and one smaller blastomere that produces a mesoderm, endoderm, and finally a germ line in a series of asymmetric divisions. This early embryonic lineage provides a model for asymmetric stem cell divisions because each division produces one daughter cell that will produce only somatic cells and a second daughter cell that is capable of generating a germ line [129–132]. These embryonic divisions occur according to mechanisms that are widely used by asymmetrically dividing stem cells and progenitors. Asymmetric division of *C. elegans* zygotes requires asymmetric localization of the PAR-3, PAR-6 and atypical protein kinase C (PAR–aPKC) complex at the cortex [140]. The asymmetrically localized PAR proteins in turn govern both mitotic spindle orientation and asymmetric segregation of cytoplasmic cell fate determinants, including riboprotein particles known as P-granules and PIE-1, a transcriptional repressor required for germline fate [140–144]. Asymmetric division of the *Drosophila* neuroblast is similarly controlled by a related mechanism [131, 145] where in an evolutionarily conserved cell fate determinant, Numb, is asymmetrically localized to daughter cells that are destined to differentiate [146]. A classic example of an asymmetric division that is controlled by an extrinsic mechanism is provided by the *Drosophila* germline stem cell. They divide with a reproducible orientation to generate one daughter that remains in the stem cell niche and retains stem cell identity, and one daughter that is placed away from the niche and begins to differentiate [132, 147, 148]. A stem cell niche is defined as a 'microenvironment' that promotes stem cell maintenance.

Cells that create stem cell niches include cap cells in the *Drosophila* ovary and hub cells in the *Drosophila* testis. Cap cells synthesize ligands called decapentaplegic (DPP) and glass bottom boat (GBB) that activate bone morphogenetic protein (BMP) signaling in germline stem cells, thereby repressing the gene bag-of-marbles, which encodes a protein that promotes differentiation. Hub cells synthesize a ligand called unpaired that activates the JAK–STAT (Janus kinase and signal transducer and activator of transcription) signaling pathway in germline stem cells to prevent differentiation by controlling target genes [122, 149–154]. Specialized junctions at the interface between the niche and germline stem cells anchor the stem cell to the niche [155, 156]. The mechanism controlling orientation of the mitotic spindle relies on centrosomal components in spermatogonial stem cells [156]. The orientation of these asymmetric stem cell divisions controls the location of daughter cells and thus their access to extrinsic signals that regulate stem cell identity. Asymmetric divisions can be controlled by both intrinsic partitioning of fate regulators and asymmetric exposure to extrinsic activators [139].

4.5 Differentiation programs

The regulation of stem cell differentiation is a challenge in regenerative medicine. Close cell–cell interactions and three-dimensional culture conditions are often necessary prerequisites for differentiation and cells within three-dimensional culture display distinct features that are more representative of native tissues than are cells within conventional two-dimensional culture [147, 151, 157–159]. Tissue engineering (TE) seeks to repair or regenerate damaged or diseased tissue and organs through the implantation of combinations of cells, scaffolds, and soluble mediators [160]. TE offers the advantage of a three-dimensional environment, as well as flexibility of size and shape and increasingly controllable environmental conditions that can be studied in depth *in vitro*. Deciphering the cellular responses to exogenous stimuli is important to understand and control tissue development and remodeling for the generation of optimal tissue engineered grafts. For a stem cell to commit to a particular program is highly contextual and requires multiple targets in different pathways to be simultaneously activated to generate a cellular response switching between growth, differentiation, and apoptosis. Multipotency makes stem cells promising cell sources for regenerative medicine therapies. However, multipotency can also lead to unwanted differentiation of an undesired cell type at an unwanted location or time which may have a detrimental effect on the native physiologic state. Stem cells have developed mechanisms and checkpoints that ensure a differentiation response only when activation occurs in the appropriate biological context. A number of factors play a role in stem cell differentiation including, growth factors and cytokines, cell–cell contacts, cell–ECM contacts, and physical forces [161].

During development, cells of the embryo make a number of cell fate decisions that result in decreasing potency eventually producing specialized cells. Cells of the embryo make cell fate decisions based on a number of chemical and physical signals that are produced within the embryo, including: secreted chemicals from neighboring cells, direct communication with neighboring cells by cell–cell contact, or mechanical strain caused by movements of the embryo as it develops. All of these processes activate signaling that ultimately leads to changes in gene activity, which determines cellular differentiation. Stem cell biologists induce stem cells to differentiate into specialized cell types by recreating an embryonic environment in a petri dish. To recreate embryonic development in a dish, stem cells are exposed to chemical or physical signals that will cause cellular differentiation, including the following:

- *Growth factors* are signaling molecules that induce cellular differentiation when added to stem cell culture medium to transform stem cells into specialized cell types.
- *Cell culture substrate* includes ECM proteins on top of which stem cells are cultured to cause differentiation.
- *Co-culture environments* can induce differentiation when stem cells are cultured with other cells that produce signaling molecules. For example, stem cells can be grown together with other cells that produce growth factors that will cause differentiation.
- *Three-dimensional cultures* include three-dimensional aggregates or spherical clumps of cells called embryoid bodies to create mechanical stimulation and cell contact and recreate an early developmental process called gastrulation that establishes the three primary germ layers.
- *Signal inhibition* occurs when inhibitory growth factors are added to inhibit the differentiation of stem cells.

The production of pancreas cells from pluripotent stem cells is an example of stem cell differentiation using these tools. Stem cells are grown as embryoid bodies together with pancreas-inducing growth factors, separated into single cells and then grown in a two-dimensional environment with another set of growth factors that will drive differentiation to mature pancreas cells. Scientific proof of differentiated cells lies in a number of criteria including:

- *Appearance or morphology* unique to a specific cell type. Cells are unique in shape depending on the function they perform, and imaging techniques are employed to classify them according to their shape and size.
- *Behavior:* specialized cells perform specialized functions. For example, heart muscle cells contract and relax and in the laboratory a differentiated cell can respond to electrical signals for the cue to contract. In the laboratory, the ability of heart cells to respond to electrical signals can be tested.
- *Safety:* if specialized cells generated from stem cells are to be used to treat human disease, they must be fully differentiated and have lost their ability to generate any other cell type [162].

4.6 Stem cell therapies

Stem cell therapies bring substantial benefit to patients suffering a wide range of diseases and injuries. A significant investment of time and resources has been made in the field of stem cell research and clinical trials. Successful new therapies come at a considerable cost that cannot easily be sustained without evaluation and guidance to enable an understanding of the extended timeframes involved and the collateral losses for potential products that are unable to meet the demands of the regulatory system and clinical efficacy of therapy. The data of many trials are unavailable or not easily accessed and this emphasizes the importance of publishing early clinical trials [163].

Autologous endothelial stem or progenitor cells show promising results for vascular disease and deficiency, and these cells in combination with MSCs and pharmaceutical intervention may be shown to be effective. Placental, BM, and fat-derived stromal cells or MSCs are currently used in the largest number of clinical trials and have been shown to be safe following transplantation. Since these cells are immune suppressive it may be that they are rapidly turned over in tissues such as the lung where they collect in large numbers. In addition, there seems to be very little difference between autologous and allogeneic MSCs in their actions or clinical effects. This suggests that they induce activation of cytokines that have immune-modulatory effects on endogenous tissue regeneration. They migrate to inflammatory sites and have inflammatory suppressing effects that are pro-regenerative for the affected tissue. Although they are not really identified as colonizing in tissue repair mechanisms, there is a question regarding their actual mechanism of action. Clinical benefits have been observed in many early clinical trials that have attracted continued funding for larger-scale studies. The contributions of bone MSCs to bone and cartilage repair and reduction of osteoarthritic and lower-back pain are impressive. MSCs from sources other than BM have not been shown to have clinical benefits and it may be premature to imply that placental stem cells have a more significant role in clinical applications. Failures in clinical trials may be predicted where insufficient scientific data support a strong clinical benefit. Clinical benefit in phase II studies must be significant to warrant phase III or IV studies, since the diversity of human disease may negate the significance of minor benefits apparent in early trials. However, if the need for regulated phase III studies can be dispensed with, many more products may become available in a shorter timeframe and more cost-effective manner. Cell therapies are rapidly evolving but few at present would have demonstrated sufficient clinical benefit to warrant their adoption as useful therapeutic modalities in a regulatory system. Scientific evidence of clinical benefit and mechanisms of action in preclinical trials will improve current controversy surrounding the safety and efficiency of this treatment, leading to a significant increase in stem cell products that will meet the criteria for registered products in the established regulatory systems [163].

References

[1] Zhu Z and Huangfu D 2013 Human pluripotent stem cells: an emerging model in developmental biology *Development* **140** 705–717

[2] ISSCR 2015 Types of stem cells. Stem cell facts *A Closer Look at Stem Cells* available from www.closerlookatstemcells.org/learn-about-stem-cells/types-of-stem-cells

[3] Lavker R R and Sun T T 2000 Epidermal stem cells: properties, markers and location *Proc. Natl Acad. Sci. USA* **97** 13473

[4] Uchida N, Buck D W, He D, Reitsma M J, Masek M, Phan T V, Tsukamoto A S, Gage F H and Weissman I L 2000 Direct isolation of human central nervous system stem cells *Proc. Natl Acad. Sci. USA* **97** 14720

[5] Vessey C J and de la Hall P M 2001 Hepatic stem cells: a review *Pathology* **33** 130

[6] Wagers A J, Christensen J L and Weissman I L 2002 Cell fate determination from stem cells *Gene Ther.* **9** 606

[7] Blau H M, Brazelton T R and Weimann J M 2001 The evolving concept of a stem cell: entity or function? *Cell* **105** 829

[8] Morrison S J 2001 Stem cell potential: can anything make anything? *Curr. Biol.* **11** R7

[9] Prockop D J, Gregory C A and Spees J L 2003 One strategy for cell and gene therapy: harnessing the power of adult stem cells to repair tissues *Proc. Natl Acad. Sci. USA* **100** 11917

[10] Finch B W and Ephrussii B 1967 Retention of multiple developmental potentialities by cells of a mouse testicular teratocarcinoma during prolonged culture *in vitro* and their extinction upon hybridization with cells of permanent lines *Proc. Natl Acad. Sci. USA* **57** 615

[11] Andrews P W 2002 From teratocarcinomas to embryonic stem cells *Philos. Trans. R. Soc. Lond. B* **357** 405

[12] Orive G, Hernandez R M, Gascon A R, Igartua M and Pedraz J L 2003 Controversies over stem cell research *Trends Biotechnol.* **21** 109

[13] Rippon H J and Bishop A E 2004 Embryonic stem cells *Cell Prolif.* **37** 23–34

[14] NIH 2017 Stem cell information. Stem cell basics IV *National Institutes of Health* available from https://stemcells.nih.gov/info/basics/4.htm

[15] Thomson J A, Itskovitz-Eldor J, Shapiro S S, Waknitz M A, Swiergiel J J, Marshall V S and Jones J M 1998 Embryonic stem cell lines derived from human blastocysts *Science* **282** 1145–147

[16] Takahashi K and Yamanaka S 2006 Induction of pluripotent stem cells from mouse embryonic and adult fibroblast cultures by defined factors *Cell* **126** 663–76

[17] Takahashi K, Tanabe K, Ohnuki M, Narita M, Ichisaka T, Tomoda K and Yamanaka S 2007 Induction of pluripotent stem cells from adult human fibroblasts by defined factors *Cell* **131** 861–72

[18] Yu J *et al* 2007 Induced pluripotent stem cell lines derived from human somatic cells *Science* **318** 1917–20

[19] Wu S M and Hochedlinger K 2011 Harnessing the potential of induced pluripotent stem cells for regenerative medicine *Nat. Cell Biol.* **13** 497–505

[20] Guenther M G *et al* 2005 Core transcriptional regulatory circuitry in human embryonic stem cells *Cell* **122** 947–56

[21] Bai X *et al* 2010 TIF1gamma controls erythroid cell fate by regulating transcription elongation *Cell* **142** 133–143

[22] Zeitlinger J, Stark A, Kellis M, Hong J W, Nechaev S, Adelman K, Levine M and Young R A 2007 RNA polymerase stalling at developmental control genes in the *Drosophila melanogaster* embryo *Nat. Genet.* **39** 1512–16

[23] Zhang X *et al* 2010 Pax6 is a human neuroectoderm cell fate determinant *Cell Stem Cell* **7** 90–100

[24] Wandzioch E and Zaret K S 2009 Dynamic signaling network for the specification of embryonic pancreas and liver progenitors *Science* **324** 1707–10

[25] Nostro M C *et al* 2011 Stage specific signaling through TGF family members and WNT regulates patterning and pancreatic specification of human pluripotent stem cells *Development* **138** 861–71

[26] Rezania A, Riedel M J, Wideman R D, Karanu F, Ao Z, Warnock G L and Kieffer T J 2011 Production of functional glucagon-secreting α-cells from human embryonic stem cells *Diabetes* **60** 239–47

[27] Zhao W, Ji X, Zhang F, Li L and Ma L 2012 Embryonic stem cell markers *Molecules* **17** 6196-236

[28] Prowse A B, McQuade L R, Bryant K J, Marcal H and Gray P P 2007 Identification of potential pluripotency determinants for human embryonic stem cells following proteomic analysis of human and mouse fibroblast conditioned media *J. Proteome Res.* **6** 3796–3807

[29] Tárnok A, Ulrich H and Bocsi J 2010 Phenotypes of stem cells from diverse origin *Cytometry* A **77** 6–10

[30] Scholer H R, Hatzopoulos A K, Balling R, Suzuki N and Gruss P 1989 A family of octamerspecific proteins present during mouse embryogenesis—evidence for germline-specific expression of an oct factor *EMBO J.* **8** 2543–50

[31] Shamblott M J, Axelman J, Wang S, Bugg E M, Littlefield J W, Donovan P J, Blumenthal P D, Huggins G R and Gearhart J D 1998 Derivation of pluripotent stem cells from cultured human primordial germ cells *Proc. Natl Acad. Sci. USA* **95** 13726–31

[32] Fox N, Damjanov I, Martinez-Hernandez A, Knowles B B and Solter D 1981 Immunohistochemical localization of the early embryonic antigen (ssea-1) in postimplantation mouse embryos and fetal and adult tissues *Dev. Biol.* **83** 391–8

[33] Fox N, Shevinsky L, Knowles B B, Solter D and Dawjanov I 1982 Distribution of murine stage specific embryonic antigens in the kidneys of three rodent species *Exp. Cell Res.* **140** 331–9

[34] Henderson J K, Draper J S, Baillie H S, Fishel S, Thomson J A, Moore H and Andrews P W 2002 Preimplantation human embryos and embryonic stem cells show comparable expression of stage-specific embryonic antigens *Stem Cells* **20** 329–37

[35] Kannagi R, Cochran N A, Ishigami F, Hakomori S, Andrews P W, Knowles B B and Solter D 1983 Stage-specific embryonic antigens (ssea-3 and -4) are epitopes of a unique globo-series ganglioside isolated from human teratocarcinoma cells *EMBO J.* **2** 2355–61

[36] Knowles B B, Aden D P and Solter D 1978 Monoclonal antibody detecting a stage-specific embryonic antigen (ssea-1) on preimplantation mouse embryos and teratocarcinoma cells *Curr. Top. Microbiol. Immunol.* **81** 51–53

[37] Shevinsky L H, Knowles B B, Damjanov I and Solter D 1982 Monoclonal antibody to murine embryos defines a stage-specific embryonic antigen expressed on mouse embryos and human teratocarcinoma cells *Cell* **30** 697–705

[38] Adewumi O *et al* 2007 Characterization of human embryonic stem cell lines by the international stem cell initiative *Nat. Biotechnol.* **25** 803–16

[39] Assou S *et al* 2007 A meta-analysis of human embryonic stem cells transcriptome integrated into a web-based expression atlas *Stem Cells* **25** 961–73

[40] Bhattacharya B *et al* 2004 Gene expression in human embryonic stem cell lines: unique molecular signature *Blood* **103** 2956–64

[41] Draper J S, Pigott C, Thomson J A and Andrews P W 2002 Surface antigens of human embryonic stem cells: changes upon differentiation in culture *J. Anat.* **200** 249–58

[42] Lian Q Z *et al* 2007 Derivation of clinically compliant MSCs from CD105+, CD24-differentiated human ESCs *Stem Cells* **25** 425–36

[43] Puri R K, Bhattacharya B, Miura T, Mejido J, Luo Y Q, Yang A X, Joshi B H, Irene G and Rao M 2004 Microrray analysis of gene expression identities unique molecular signature in human embryonic stem cell lines *FASEB J.* **18** A1121

[44] Skottman H, Mikkola M, Lundin K, Olsson C, Stromberg A M, Tuuri T, Otonkoski T, Hovatta O and Lahesmaa R 2005 Gene expression signatures of seven individual human embryonic stem cell lines *Stem Cells* **23** 1343–56

[45] Sundberg M, Jansson L, Ketolainen J, Pihlajamaki H, Suuronen R, Skottman H, Inzunza J, Hovatta O and Narkilahti S 2009 Cd marker expression profiles of human embryonic stem cells and their neural derivatives, determined using flow-cytometric analysis, reveal a novel cd marker for exclusion of pluripotent stem cells *Stem Cell Res.* **2** 113–24

[46] Xu C H, Inokuma M S, Denham J, Golds K, Kundu P, Gold J D and Carpenter M K 2001 Feeder-free growth of undifferentiated human embryonic stem cells *Nat. Biotechnol.* **19** 971–74

[47] Yin A H, Miraglia S, Zanjani E D, AlmeidaPorada G, Ogawa M, Leary A G, Olweus J, Kearney J and Buck D W (1997) AC133, a novel marker for human hematopoietic stem and progenitor cells. *Blood* **90** 5002–12

[48] Aumailley M, Timpl R and Sonnenberg A 1990 Antibody to integrin alpha 6 subunit specifically inhibits cell-binding to laminin fragment 8. Exp. *Cell Res.* **188** 55–60

[49] Chute J P 2006 Stem cell homing *Curr. Opin. Hematol.* **13** 399–406

[50] Fassler R, Pfaff M, Murphy J, Noegel A A, Johansson S, Timpl R and Albrecht R 1995 Lack of beta 1 integrin gene in embryonic stem cells affects morphology, adhesion, and migration but not integration into the inner cell mass of blastocysts *J. Cell Biol.* **128** 979–88

[51] Hall D E, Reichardt L F, Crowley E, Holley B, Moezzi H, Sonnenberg A and Damsky C H 1990 The alpha 1/beta 1 and alpha 6/beta 1 integrin heterodimers mediate cell attachment to distinct sites on laminin *J. Cell Biol.* **110** 2175–84

[52] Harris E S, McIntyre T M, Prescott S M and Zimmerman G A 2000 The leukocyte integrins *J. Biol. Chem.* **275** 23409–412

[53] Lee S T, Yun J I, Jo Y S, Mochizuki M, van der Vlies A J, Kontos S, Ihm J E, Lim J M and Hubbell J A 2010 Engineering integrin signaling for promoting embryonic stem cell self-renewal in a precisely defined niche *Biomaterials* **31** 1219–26

[54] Qian H, Georges-Labouesse E, Nystrom A, Domogatskaya A, Tryggvason K, Jacobsen S E and Ekblom M 2007 Distinct roles of integrins alpha6 and alpha4 in homing of fetal liver hematopoietic stem and progenitor cells *Blood* **110** 2399–2407

[55] Rabinovitz I, Nagle R B and Cress A E 1995 Integrin alpha-6 expression in human prostate carcinoma-cells is associated with a migratory and invasive phenotype *in vitro* and *in vivo* *Clin. Exp. Metastasis* **13** 481–91

[56] Ruoslahti E and Pierschbacher M D 1987 New perspectives in cell adhesion: RGD and integrins *Science* **238** 491–97

[57] Watt F M and Hogan B L 2000 Out of eden: Stem cells and their niches *Science* 287 1427–30

[58] Andrews P W, Banting G, Damjanov I, Arnaud D and Avner P 1984 Three monoclonal antibodies defining distinct differentiation antigens associated with different high molecular weight polypeptides on the surface of human embryonal carcinoma cells *Hybridoma* **3** 347–61

[59] Giwercman A, Andrews P W, Jorgensen N, Muller J, Graem N and Skakkebaek N E 1993 Immunohistochemical expression of embryonal marker TRA-1–60 in carcinoma *in situ* and germ cell tumors of the testis *Cancer* **72** 1308–14

[60] Schopperle W M and DeWolf W C 2007 The TRA-1–60 and TRA-1–81 human pluripotent stem cell markers are expressed on podocalyxin in embryonal carcinoma *Stem Cells* **25** 723–30

[61] Barker N and Clevers H   2000) Catenins, WNT signaling and cancer. *Bioessays* **22**   961–65

[62] Katoh Y and Katoh M 2007 Conserved POU-binding site linked to SP1-binding site within FZD5 promoter: transcriptional mechanisms of FZD5 in undifferentiated human ES cells,

fetal liver/spleen, adult colon, pancreatic islet, and diffuse-type gastric cancer. *Int. J. Oncol.* **30** 751–55

[63] Layden B T, Newman M, Chen F, Fisher A and Lowe W L 2010 G protein coupled receptors in embryonic stem cells: a role for GS-alpha signaling. *PLoS One* **5** e9105

[64] Okoye U C, Malbon C C and Wang H Y 2008 WNT and frizzled RNA expression in human mesenchymal and embryonic (H7) stem cells. *J. Mol. Signal.* **3** 16

[65] Polakis P 2000 Wnt signaling and cancer *Genes Dev.* **14** 1837–51

[66] Geissler E N, Liao M, Brook J D, Martin F H, Zsebo K M, Housman D E and Galli S J 1991 Stem cell factor (SCF), a novel hematopoietic growth factor and ligand for c-Kit tyrosine kinase receptor, maps on human chromosome 12 between 12q14.3 and 12ter *Somat. Cell Mol. Genet.* **17** 207–14

[67] Bashamboo A, Taylor A H, Samuel K, Panthier J J, Whetton A D and Forrester L M 2006 The survival of differentiating embryonic stem cells is dependent on the SCF–Kit pathway *J. Cell Sci.* **119** 3039–46

[68] Lonardo E, Parish C L, Ponticelli S, Marasco D, Ribeiro D, Ruvo M, De Falco S, Arenas E and Minchiotti G 2010 A small synthetic cripto blocking peptide improves neural induction, dopaminergic differentiation, and functional integration of mouse embryonic stem cells in a rat model of Parkinson's disease *Stem Cells* **28** 1326–37

[69] Strohmeyer T, Reese D, Press M, Ackermann R, Hartmann M and Slamon D 1995 Expression of the c-Kit proto-oncogene and its ligand stem cell factor (SCF) in normal and malignant human testicular tissue *J. Urol.* **153** 511–15

[70] Pesce M and Scholer H R 2000 Oct-4: control of totipotency and germline determination *Molecular Reprod. Dev.* **55** 452–57

[71] Pesce M and Scholer H R 2001 Oct-4: gatekeeper in the beginnings of mammalian development *Stem Cells* **19** 271–78

[72] Botquin V, Hess H, Fuhrmann G, Anastassiadis C, Gross M K, Vriend G and Scholer H R 1998 New POU dimer configuration mediates antagonistic control of an osteopontin preimplantation enhancer by Oct-4 and Sox-2 *Genes Dev.* **12** 2073–90

[73] Boyer L A *et al* 2005 Core transcriptional regulatory circuitry in human embryonic stem cells *Cell* **122** 947–56

[74] Jiang J M, Chan Y S, Loh Y H, Cai J, Tong G Q, Lim C A, Robson P, Zhong S and Ng H H 2008 A core KLF circuitry regulates self-renewal of embryonic stem cells *Nat. Cell Biol.* **10** 353–60

[75] Chambers I, Colby D, Robertson M, Nichols J, Lee S, Tweedie S and Smith A 2003 Functional expression cloning of nanog, a pluripotency sustaining factor in embryonic stem cells *Cell* **113** 643–55

[76] Hatano S, Tada M, Kimura H, Yamaguchi S, Kono T, Nakano T, Suemori H, Nakatsuji N and Tada T 2005 Pluripotential competence of cells associated with nanog activity *Mech. Dev.* **122** 67–79

[77] Mitsui K, Tokuzawa Y, Itoh H, Segawa K, Murakami M, Takahashi K, Maruyama M, Maeda M and Yamanaka S 2003 The homeoprotein nanog is required for maintenance of pluripotency in mouse epiblast and ES cells *Cell* **113** 631–42

[78] Gordon S, Akopyan G, Garban H and Bonavida B 2006 Transcription factor yy1: structure, function, and therapeutic implications in cancer biology. *Oncogene* **25** 1125–42

[79] Koestenbauer S, Zech N H, Juch H, Vanderzwalmen P, Schoonjans L and Dohr G 2006 Embryonic stem cells: Similarities and differences between human and murine embryonic stem cells *Am. J. Reprod. Immunol.* **55** 169–80

[80] Rogers M B, Hosler B A and Gudas L J 1991 Specific expression of a retinoic acid-regulated, zinc-finger gene, Rex-1, in preimplantation embryos, trophoblast and spermatocytes Development **113** 815–24

[81] Kooistra S M and Thummer R P; Eggen B J 2009 Characterization of human UTF1, a chromatin associated protein with repressor activity expressed in pluripotent cells. *Stem Cell Res.* **2** 211–18

[82] van den Boom, V, Kooistra S M, Boesjes M, Geverts B, Houtsmuller A B, Monzen K, Komuro I, Essers J, Drenth-Diephuis L J and Eggen B J (2007) UTF1 is a chromatin-associated protein involved in ES cell differentiation. *J. Cell Biol.* **178** 913–24

[83] Galan-Caridad J M, Harel S, Arenzana T L, Hou Z E, Doetsch F K, Mirny L A and Reizis B (2007) ZFX controls the self-renewal of embryonic and hematopoietic stem cells. *Cell* **129** 345–57

[84] Kopito R R, Lee B S, Simmons D M, Lindsey A E, Morgans C W and Schneider K 1989 Regulation of intracellular pH by a neuronal homolog of the erythrocyte anion-exchanger. *Cell* **59** 927–37

[85] Voss A K, Thomas T, Petrou P, Anastassiadis K, Scholer H and Gruss P 2000 Taube nuss is a novel gene essential for the survival of pluripotent cells of early mouse embryos Development **127** 5449–61

[86] Liu Y and Labosky P A 2008 Regulation of embryonic stem cell self-renewal and pluripotency by *Stem Cells* **26** 2475–84

[87] Momma T, Hanna L A, Clegg M S and Keen C L 2002 Zinc influences the *in vitro* development of peri-implantation mouse embryos *FASEB J.* **16** A652

[88] Sutton J *et al* 1996 Genesis, a winged helix transcriptional repressor with expression restricted to embryonic stem cells *J. Biol. Chem.* **271** 23126–133

[89] Li H, Fu X, Ouyang Y, Cai C L, Wang J and Sun T 2006 Adult bone-marrow derived mesenchymal stem cells contribute to wound healing of skin appendages *Cell Tissue Res* **326** 725– 36

[90] Monzen K *et al* 2008 A crucial role of a high mobility group protein HMGA2 in cardiogenesis *Nat. Cell Biol.* **10** 567–74

[91] Nishino J, Kim I, Chada K and Morrison S J 2008 HMGA2 promotes neural stem cell self-renewal in young but not old mice by reducing p16Ink4a and p19Arf expression *Cell* **135** 227–39

[92] Pfannkuche K, Summer H, Li O, Hescheler J and Droge P 2009 The high mobility group protein HMGA2: a co-regulator of chromatin structure and pluripotency in stem cells? *Stem Cell Rev.* **5** 224–30

[93] Kalivas P W, Duffy P and Mackler S A 1999 Interrupted expression of NAC-1 augments the behavioural responses to cocaine *Synapse* **33** 153–59

[94] Kim Y M, Jeon E S, Kim M R, Jho S K, Ryu S W and Kim J H 2008 Angiotensin ii-induced differentiation of adipose tissue-derived mesenchymal stem cells to smooth muscle like cells *Int. J. Biochem. Cell Biol.* **40** 2482–91

[95] Mackler S A, Korutla L, Cha X Y, Koebbe M J, Fournier K M, Bowers M S and Kalivas P W 2000 NAC-1 is a brain POZ/BTB protein that can prevent cocaine-induced sensitization in the rat *J. Neurosci.* **20** 6210–17

[96] Lan Z J, Xu X, Chung A C and Cooney A J 2009 Extra-germ cell expression of mouse nuclear receptor subfamily 6, group a, member 1 (NR6A1) *Biol. Reprod* **80** 905–12

[97] Lei W, Hirose T, Zhang L X, Adachi H, Spinella M J, Dmitrovsky E and Jetten A M 1997 Cloning of the human orphan receptor germ cell nuclear factor/retinoid receptor-related testis-associated receptor and its differential regulation during embryonal carcinoma cell differentiation *J. Mol. Endocrinol.* **18** 167–76

[98] Heim M H 1999 The Jak-STAT pathway: cytokine signalling from the receptor to the nucleus *J. Recept. Signal Transduct. Res.* **19** 75–120

[99] Takada I *et al* 2007 A histone lysine methyltransferase activated by non-canonical Wnt signalling suppresses PPAR-gamma transactivation *Nat. Cell Biol.* **9** 1273–85

[100] Schilham M W and Clevers H 1998 HMG box containing transcription factors in lymphocyte differentiation *Semin. Immunol.* **10** 127–32

[101] Kohlhase J, Pasche B, Burfeind P, Wischermann A, Reichenbach H, Froster U and Engel W 1998 Mutations in the Sall1 putative transcription factor gene cause Townes–Brocks syndrome *Eur. J. Hum. Genet.* **6** 33–33

[102] Zhang J *et al* 2006 Sall4 modulates embryonic stem cell pluripotency and early embryonic development by the transcriptional regulation of Pou5f1 *Nat. Cell Biol.* **8** 1114–23

[103] Tokuzawa Y, Kaiho E, Maruyama M, Takahashi K, Mitsui K, Maeda M, Niwa H and Yamanaka S 2003 Fbx15 is a novel target of Oct3/4 but is dispensable for embryonic stem cell self-renewal and mouse development *Mol. Cell. Biol.* **23** 2699–708

[104] Wong R C B, Ibrahim A, Fong H, Thompson N, Lock L F and Donovan P J 2011 L1TD1 is a marker for undifferentiated human embryonic stem cells *PLoS One* **6** e19355

[105] Pierre A, Gautier M, Callebaut I, Bontoux M, Jeanpierre E, Pontarotti P and Monget P 2007 Atypical structure and phylogenomic evolution of the new eutherian oocyte-and embryoexpressed KHDC1/DPPA5/ECAT1/OOEP gene family *Genomics* **90** 583–94

[106] Levine A J and Brivanlou A H 2006 GDF3, a BMP inhibitor, regulates cell fate in stem cells and early embryos *Development* **133** 209–16

[107] Tanaka T S, de Silanes I L, Sharova L V, Akutsu H, Yoshikawa T, Amano H, Yamanaka S, Gorospe M and Ko M S H 2006 Esg1, expressed exclusively in preimplantation embryos, germline, and embryonic stem cells, is a putative RNA-binding protein with broad RNA targets *Dev. Growth Differ.* **48** 381–90

[108] Western P, Maldonado-Saldivia J, Van den Bergen, Hajkova J P, Saitou M, Barton S and Surani M A 2005 Analysis of Esg1 expression in pluripotent cells and the germline reveals similarities with Oct4 and Sox2 and differences between human pluripotent cell lines. *Stem Cells* **23** 1436–42

[109] Masaki H, Nishida T, Kitajima S, Asahina K and Teraoka H 2007 Developmental pluripotencyassociated 4 (Dppa4) localized in active chromatin inhibits mouse embryonic stem cell differentiation into a primitive ectoderm lineage *J. Biol. Chem.* **282** 33034–42

[110] Du J, Chen T J, Zou X, Xiong B and Lu G X 2010 Dppa2 knockdown-induced differentiation and repressed proliferation of mouse embryonic stem cells *J. Biochem.* **147** 265–71

[111] Maldonado-Saldivia J, den Bergen Van, Krouskos J, Gilchrist M, Lee M, Li R, Sinclair A H, Surani M A and Western P S 2007 Dppa2 and Dppa4 are closely linked SAP motif genes restricted to pluripotent cells and the germ line. *Stem Cells* **25** 19–28

[112] Bortvin A, Goodheart M, Liao M and Page D C 2004 Dppa3/Pgc7/stella is a maternal factor and is not required for germ cell specification in mice *BMC Dev. Biol.* **4** 2

[113] Bowles J, Teasdale R P, James K and Koopman P 2003 Dppa3 is a marker of pluripotency and has a human homologue that is expressed in germ cell tumours *Cytogenet. Genome Res.* **101** 261–65

[114] Miyabayashi T, Teo J L, Yamamoto M, McMillan M, Nguyen C and Kahn M 2007 Wnt/beta-catenin/CBP signaling maintains long-term murine embryonic stem cell pluripotency *Proc. Natl Acad. Sci. USA* **104** 68–73

[115] Takao Y, Yokota T and Koide H 2007 Beta-catenin up-regulates Nanog expression through interaction with Oct-3/4 in embryonic stem cells *Biochem. Biophys. Res. Commun.* **353** 699–705

[116] Datto M and Wang X F 2000 The Smads: transcriptional regulation and mouse models *Cytokine Growth Factor Rev.* **11** 37–48

[117] Feng X H and Derynck R 2005 Specificity and versatility in TGF-beta signaling through Smads *Annu. Rev. Cell Dev. Biol.* **21** 659–93

[118] Massague J, Seoane J and Wotton D 2005 Smad transcription factors *Genes Dev.* **19** 2783–810

[119] Preffer F and Dombkowski D 2009 Advances in complex multiparameter flow cytometry technology: applications in stem cell research *Cytometry* B **76** 295–314

[120] Ferraro F and Lo Celso C 2010 Adult stem cells and their niches *Adv Exp Med Biol* **695** 155–68

[121] Terskikh V V, Vasiliev A V and Vorotelyak E A 2007 Stem cell niches *Biol. Bull. Russ. Acad. Sci.* **34** 211

[122] Schofield R 1978 The relationship between the spleen colony-forming cell and the haemopoietic stem cell *Blood Cells* **4** 7–25

[123] Xie T and Spradling A C 1998 Decapentaplegic is essential for the maintenance and division of germline stem cells in the *Drosophila* ovary *Cell* **94** 251–60

[124] Kimble J E and White J G 1981 On the control of germ cell development in Caenorhabditis elegans *Dev. Biol.* **81** 208–19

[125] Katayama Y *et al* 2006 Signals from the sympathetic nervous system regulate hematopoietic stem cell egress from bone marrow *Cell* **124** 407–21

[126] Mayack S R *et al* 2010 Systemic signals regulate ageing and rejuvenation of blood stem cell niches *Nature* **463** 495–500

[127] Conboy I M *et al* 2005 Rejuvenation of aged progenitor cells by exposure to a young systemic environment *Nature* **433** 760–64

[128] Knoblich J A 2008 Mechanisms of asymmetric stem cell division *Cell* **132** 583–597

[129] Betschinger J and Knoblich J A 2004 Dare to be different: asymmetric cell division in *Drosophila*, *C. elegans* and vertebrates *Curr. Biol.* **14** R674–85

[130] Clevers H 2005 Stem cells, asymmetric division and cancer *Nature Genet.* **37** 1027–28

[131] Doe C Q and Bowerman B 2001 Asymmetric cell division: fly neuroblast meets worm zygote *Curr. Opin. Cell Biol.* **13** 68–75

[132] Yamashita Y M, Fuller M T and Jones D L 2005 Signaling in stem cell niches: lessons from the *Drosophila* germline *J. Cell Sci.* **118** 665–72

[133] Morrison S J, Hemmati H D, Wandycz A M and Weissman I L 1995 The purification and characterization of fetal liver hematopoietic stem cells *Proc. Natl. Acad. Sci. USA* **92** 10302–306

[134] Lechler T and Fuchs E 2005 Asymmetric cell divisions promote stratification and differentiation of mammalian skin *Nature* **437** 275–80

[135] Wright D E *et al* 2001 Cyclophosphamide/granulocyte colony-stimulating factor causes selective mobilization of bone marrow hematopoietic stem cells into the blood after M phase of the cell cycle *Blood* **97** 2278–85

[136] Morrison S J, Wright D and Weissman I L 1997 Cyclophosphamide/granulocyte colony stimulating factor induces hematopoietic stem cells to proliferate prior to mobilization *Proc. Natl Acad. Sci. USA* **94** 1908–13

[137] Bodine D, Seidel N E and Orlic D 1996 Bone marrow collected 14 days after *in vivo* administration of granulocyte colony-stimulating factor and stem cell factor to mice has 10-fold more repopulating ability than untreated bone marrow *Blood* **88** 89–97

[138] Doetsch F, Petreanu L, Caille I, Garcia-Verdugo J M and Alvarez-Buylla A 2002 EGF converts transit-amplifying neurogenic precursors in the adult brain into multipotent stem cells *Neuron* **36** 1021–34

[139] Morrison S J and Kimble J 2006 Asymmetric and symmetric stem-cell divisions in development and cancer *Nature* **441** 1068–74

[140] Gönczy P and Rose L S 2005 *Asymmetric cell division and axis formation in the embryo* WormBook *ed The* C. elegans *Research Community* available from www.wormbook.org/chapters/www_asymcelldiv/asymcelldiv.html

[141] Strome S and Wood W B 1983 Generation of asymmetry and segregation of germ-line granules in early *C. elegans* embryos *Cell* **35** 15–25

[142] Mello C C, Draper B W, Krause M, Weintraub H and Priess J R 1992 The pie-1 and mex-1 genes and maternal control of blastomere identity in early *C. elegans* embryos *Cell* **70** 163–76

[143] Mello C C *et al* 1996 The PIE-1 protein and germline specification in *C. elegans* embryos *Nature* **382** 710–12

[144] Reese K J, Dunn M A, Waddle J A and Seydoux G 2000 Asymmetric segregation of PIE-1 in *C. elegans* is mediated by two complementary mechanisms that act through separate PIE-1 protein domains *Mol. Cell* **6** 445–55

[145] Wodarz A 2005 Molecular control of cell polarity and asymmetric cell division in *Drosophila* neuroblasts *Curr. Opin. Cell Biol.* **17** 475–81

[146] Spana E P, Kopczynski C, Goodman C S and Doe C Q 1995 Asymmetric localization of numb autonomously determines sibling neuron identity in the *Drosophila* CNS *Development* **121** 3489–94

[147] Xie T and Spradling A C 2000 A niche maintaining germ line stem cells in the *Drosophila* ovary *Science* **290** 328–30

[148] Spradling A, Drummond-Barbosa D and Kai T 2001 Stem cells find their niche *Nature* **414** 98–104

[149] Li L and Xie T 2005 Stem cell niche: structure and function Annu. Rev. Cell Dev. Biol. **21** 605–31

[150] Tulina N and Matunis E 2001 Control of stem cell self-renewal in *Drosophila* spermato-genesis by JAK–STAT signaling *Science* **294** 2546–49

[151] Kiger A A, Jones D L, Schulz C, Rogers M B and Fuller M T 2001 Stem cell self-renewal specified by JAK–STAT activation in response to a support cell cue *Science* **294** 2542–45

[152] Chen D and McKearin D 2003 Dpp signaling silences bam transcription directly to establish asymmetric divisions of germline stem cells *Curr. Biol.* **13** 1786–91

[153] Song X *et al* 2004 Bmp signals from niche cells directly repress transcription of a differentiation-promoting gene, bag of marbles, in germline stem cells in the *Drosophila* ovary *Development* **131** 1353–64

[154] Ohlstein B and McKearin D 1997 Ectopic expression of the *Drosophila* Bam protein eliminates oogenic germline stem cells *Development* **124** 3651–62

[155] Song X, Zhu C H, Doan C and Xie T 2002 Germline stem cells anchored by adherens junctions in the *Drosophila* ovary niches *Science* **296** 1855–57

[156] Yamashita Y M, Jones D L and Fuller M T 2003 Orientation of asymmetric stem cell divisions by the APC tumor suppressor and centrosome *Science* **301** 1547–50

[157] Abbott A 2003 Cell culture: biology's new dimension *Nature* **424** 870–72

[158] Griffith L and Swartz M 2006 Capturing complex 3D tissue physiology in vitro *Nat. Rev. Mol. Cell. Biol.* **7** 211–24

[159] Radisic M, Park H, Cerecht-Nir S, Cannizzaro C, Langer R and Vunjak-Novakovic G 2007 Biomimetic approach to cardiac tissue engineering *Philos. Trans. R. Soc. London Biol. Sci.* **362** 1357–68

[160] Guilak F, Cohen D, Estes B, Gimble J, Leidtke W and Chen C 2009 Control of Stem Cell Fate by Physical Interactions with the Extracellular Matrix *Cell Stem Cll.* **5** 17–26

[161] Clause K C, Liu L J and Tobita K 2010 Directed stem cell differentiation: the role of physical forces *Cell. Commun. Adhes.* **17** 48–54

[162] Explorecuriocity 2013 Biotechology: stem cell differentiation: stem cells *CurioCity* www. explorecuriociocity.org

[163] Trounson A and McDonald C 2015 Stem cell therapies in clinical trials: progress and challenges *Cell Stem Cell* **17** 11–22

[166] Kim J, Chu J, Shen X, Wang J and Orkin S H 2008 An extended transcriptional network for pluripotency of embryonic stem cells *Cell* **132** 1049–61

IOP Concise Physics

Photomedicine and Stem Cells
The Janus face of photodynamic therapy (PDT) to kill cancer stem cells, and photobiomodulation
(PBM) to stimulate normal stem cells
Heidi Abrahamse and Michael R Hamblin

Chapter 5

PBM and stem cells

5.1 Introduction

Stem cells are undifferentiated cells that can differentiate into more specialized cells (called progenitor cells) and can divide (through mitosis) to produce a continuous supply of stem cells. They are found in multicellular organisms, and in mammals there are two broad types of stem cells: embryonic stem cells (ESCs), which are found inside the growing cell mass of blastocysts, and adult stem cells (ASCs), which are found localized in specialized 'niches' in different tissues and organ systems. In adult organisms, stem cells and progenitor cells act as a repair system for the body, replenishing damaged, worn out, or senescent cells in adult tissues. In a developing embryo, stem cells can differentiate into all the different specialized classes of cells, giving rise to the ectoderm, endoderm, and mesoderm. Tissues that have a high rate of cell turnover, such as blood, skin, or intestines, are particularly rich in stem cells. Although ideas relevant to the concept of stem cells had been around for over a century [1], the discovery of stem cells is usually attributed to Till and McCulloch in 1963 [2]. For many years the prospect of 'stem cell therapy' for a wide range of different diseases and disorders has been seen on the horizon, but in reality has always appeared to be still out of reach [3].

ESCs that can differentiate into any cell type are called 'totipotent'; fetal stem cells have a more limited range of differentiation and are called 'pluripotent'; while ASCs are called 'multipotent', 'oligopotent', or 'unipotent' with progressively fewer choices of which cell type they can become.

Stem cells have two different mechanisms of cell division, asymmetric and stochastic symmetric [4]. Asymmetric replication involves a stem cell dividing into one cell that is identical to the original stem cell, and at the same time producing another progenitor cell that is more differentiated [5]. Stochastic symmetric differentiation means that one stem cell develops into two differentiated progenitor cells,

doi:10.1088/978-1-6817-4321-9ch5

while another different stem cell undergoes mitosis and produces two identical undifferentiated stem cells [6].

The stem cell niche is a specific anatomic location that regulates how stem cells behave. The niche protects stem cells from dying, while also protecting the host by regulating excessive stem cell proliferation [7]. The niche has both anatomical and functional attributes and integrates the signals that work together to ensure that stem cells can be made available on demand whenever there is a need to repair damaged tissue or replenish short-lived somatic cells. This control mechanism is governed by a complicated range of biological factors including adhesion molecules, extracellular matrix components, adenosine triphosphate (ATP), growth factors, and cytokines, and also physical factors such as oxygen tension, pH, and Ca^{2+} concentration in the environment.

5.2 Mechanisms of PBM/LLLT

Photobiomodulation (PBM) or low-level laser (light) therapy (LLLT) was discovered almost 50 years ago by the serendipitous action of a low-powered ruby laser in stimulating hair regrowth and wound healing in mice [8]. In the years since then it has been realized that non-coherent light from sources such as light-emitting diodes (LEDs) is also efficient to carry out PBM [9]. Moreover, although the early studies mainly used red light (600–700 nm), it was subsequently found that near-infrared (NIR) light (760–1000+ nm) was equally (if not more) effective [10]. Light in the low 700 nm regions does not appear to be particularly effective [11, 12]. This double peak in the action spectrum reflects the absorption spectrum of cytochrome c oxidase (CCO), unit IV in the mitochondrial respiratory chain [13]. Together with studies showing the effects of light on isolated mitochondria, these observations led to the formation of the most widely held hypothesis: that light stimulates respiration in mitochondria, increasing electron transport, oxygen consumption, and ATP synthesis [14]. This stimulation may happen via photodissociation of inhibitory nitric oxide from the heme and copper centers contained within CCO [15].

However, it is becoming accepted that there must be another primary photoacceptor to explain the positive effects of light at wavelengths higher than the absorption peaks of CCO, observed changes in cellular Ca^{2+}, and diverse PBM effects on nerve cells. This second photoreceptor is proposed to be a type of ion channels called 'transient receptor potential vanilloid sub-type' (TRPV) [16, 17]. TRPV calcium channels are known to be sensitive to visible light, infrared radiation, heat, cold, pressure, and spices such as capsaicin [18]. One possible explanation of this sensitivity involves the participation of 'nanostructured water' as another possible chromophore that absorbs red or NIR light [19].

Since it has been observed that positive effects of PBM *in vivo* can last a very long time (days or weeks) after a single exposure to light, investigators have begun to study the activation of signaling pathways and transcription factors that can modulate gene expression levels. Many of the factors that are known to be stimulated by light, such as nitric oxide, calcium, and reactive oxygen species

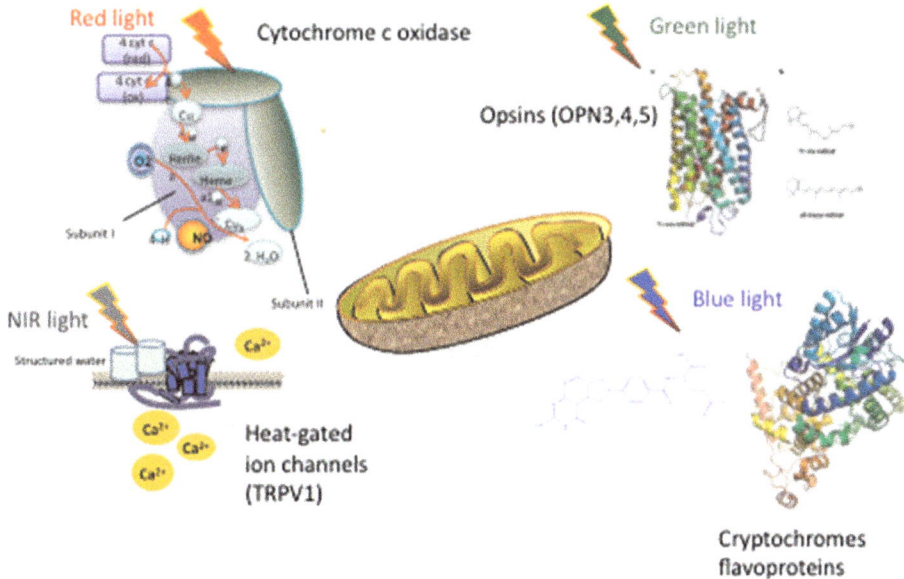

Figure 5.1. Chromophores in PBM. CCO in the respiratory chain absorbs mainly red (and NIR) light by heme and copper. Heat-gated TRP ion channels absorb NIR (and blue light) via structured water; opsins absorb mainly blue/green light via cis-retinal; and flavoproteins and cryptochromes absorb mainly blue light via pterin.

(ROS), are also known to trigger many cell-signaling pathways. Figure 5.1 shows the possible chromophores involved in the light absorption in PBM.

5.3 Why should PBM particularly affect stem cells?

As mentioned above, the stem cell niche is characterized by low oxygen tension [20]. This hypoxia is particularly evident in the endosteal zone of the bone marrow (BM) [21]. This region has a unique organization of the vascular characterized by sinusoidal capillaries, which are loosely organized and fenestrated, enabling newly formed hematopoietic cells to gain ready access to the bloodstream [22]. As a result, the perfusion of the BM is limited, and the partial pressure of oxygen (PO2) in the endosteal region is very low. The evolutionary rationale for the preference of stem cells for a markedly hypoxic niche is proposed to be that since stem cells must be able to last for the entire life-span of the organism, they need to avoid needless environmental damage. One of the most damaging agents that would affect the longevity of stem cells would be oxidative damage to DNA and other biomolecules, caused by the ROS that are an inevitable by-product of aerobic respiration (or life itself). Therefore the metabolism of stem cells tends to have an overall anaerobic character with relatively low mitochondrial activity and high expression of glycolytic enzymes [23]. The low metabolic rate of stem cells accounts for their relative quiescence and increased resistance to stress. Because stem cells must last for such a long time, they have to minimize the number of cell divisions they undergo because each cycle of division carries a small but significant risk of DNA damage.

We have formed the hypothesis that when PBM is delivered to stem cells in (for instance) the hypoxic BM, the rudimentary mitochondria in these stem cells are triggered into action and mitochondrial biogenesis can take place producing even more mitochondria [24]. Increased mitochondrial activity is accompanied by an increasing demand for oxygen, which is not so much available in the low-oxygen environment of the BM. Therefore the stem cells have to leave their niche in pursuit of the oxygen they need to satisfy their new metabolic preference involving oxidative phosphorylation. The burst of intracellular ROS that is observed to follow PBM [25] may also have a role in triggering the differentiation of stem cells [26].

5.4 Role of stem/progenitor cells in PBM/LLLT

It has been known for many years that stem cells must be involved in many of the medically beneficial applications of PBM/LLLT. It appears that stem cells are particularly sensitive to light. PBM induces stem cell activity shown by increased cell migration, differentiation, proliferation, and viability, as well as by activating protein expression. Mesenchymal stem cells, usually derived from BM, dental pulp, periodontal ligament and from adipose tissue, proliferate more after light irradiation (usually with wavelengths ranging from 600–700 nm). Since stem cells in their undifferentiated form show a lower rate of proliferation, this may be a limiting factor for the clinical effectiveness of stem cell therapies, PBM offers a viable alternative to promote the translation of stem cell research into the clinical arena.

Min and co-workers reported that the cell viability of adipose-derived stem cells was found to be increased after irradiation with 830 nm light [27]. Their *in vivo* results also revealed elevated numbers of stem cells compared to the control group. Epidermal stem cells can also be influenced by light, as demonstrated by Liao *et al* [28]. The authors reveal that 632.8 nm light has photobiological effects on cultured human epidermal stem cells, such as an increase in proliferation and migration *in vitro*. Soares observed a similar effect on human periodontal ligament stem cells irradiated with a 660 nm diode laser.

5.5 PBM to the bone marrow

Uri Oron and his group in Israel have conducted a number of studies that involve delivering light to the BM with the aim of stimulating and mobilizing stem cells. In 2011 they published a study [29] comparing PBM in rats that had had a myocardial infarction (MI, heart attack) induced by ligation of the left anterior descending coronary artery. Rats either received PBM (804 nm, 400 mW) 20 min post-MI applied directly to the myocardium of the infarcted heart or to the exposed tibia of the hind leg designed to stimulate the BM. Rats that received PBM directly to the heart had a reduction 31% in infarct size as compared to control rats, while those rats that received PBM to the BM showed a reduction of infarct size of 76%. Rats that received PBM to the heart had the same density of c-kit positive cells (mesenchymal stem cells) in the heart as controls, while those that received PBM to the BM had 25 times more c-kit cells in the heart. The group went on to replicate their results in a large animal model, farm pigs (3–4 months old, 35–40 kg) subjected to an MI by percutaneous balloon inflation in the left coronary artery for 90min [29]. PBM was applied to the tibia and iliac bones 30min, and 2 and 7 days post-induction of MI. The infarct size in the PBM-treated pigs was significantly lower, 68% than control pigs. The mean density of small blood vessels in the infarcted area and the left ventricular ejection fraction was also significantly higher. The number of c-kit+ cells (stem cells) in the circulating blood of the PBM pigs was 2.62-fold higher than in the control pigs.

Oron's group also showed that PBM to the BM could improve other diseases. They looked at a rat model of kidney damage caused by acute ischemia-reperfusion injury (IRI) [29]. Rats had one kidney surgically removed and the remaining kidney was subjected to IRI, followed by PBM to the BM (exposed tibia bones) twice (10 min and 24 h post-IRI). Histological evaluation of kidney sections revealed restored structural integrity of the renal tubules, and a significant reduction of 66% in the pathological score in the PBM-treated rats as compared to the controls. The c-kit+ cell density in kidneys post-PBM was 2.4-fold higher compared to controls. Creatinine, blood urea nitrogen, and cystatin-C levels were (significantly) 55%, 48%, and 25% lower, respectively, in the PBM rats demonstrating improved kidney function.

The Oron group also studied a transgenic mouse model of Alzheimer's disease (AD) [29]. They used 5XFAD transgenic male mice (Tg6799) that co-overexpress familial AD (FAD) mutant forms of human APP (the Swedish mutation, K670N/M671L; the Florida mutation, I716V; and the London mutation, V717I) and PS1 (M146L/L286V) trans-genes, under transcriptional control of the neuron-specific mouse Thy-1 promoter [29]. Mice received PBM to the exposed tibia six times (at 10 day intervals, for 2 months) starting at the age of 4 months (at this time the mice already had well-established AD pathology). PBM-treated mice had improved cognitive ability and spatial learning as demonstrated by the object recognition test and the fear-conditioning test. Histological examination revealed a lower burden of beta-amyloid plaque in the entire hippocampal region of the brain.

References

[1] Ramalho-Santos M and Willenbring H 2007 On the origin of the term 'stem cell' *Cell Stem Cell* **1** 35–8

[2] Becker A J, Mc C E and Till J E 1963 Cytological demonstration of the clonal nature of spleen colonies derived from transplanted mouse marrow cells *Nature* **197** 452–4

[3] Stoltz J F *et al* 2015 Stem cells and regenerative medicine: myth or reality of the 21st century *Stem Cells Int.* **2015** 734731

[4] Shahriyari L and Komarova N L 2013 Symmetric vs. asymmetric stem cell divisions: an adaptation against cancer? *PLoS One* **8** e 76195

[5] Pare J F and Sherley J L 2006 Biological principles for *ex vivo* adult stem cell expansion *Curr. Top. Dev. Biol.* **73** 141–71

[6] Sanchez-Taltavull D and Alarcon T 2014 Robustness of differentiation cascades with symmetric stem cell division *J. R. Soc. Interface* **11** 20140264

[7] Scadden D T 2014 Nice neighborhood: emerging concepts of the stem cell niche *Cell* **157** 41–50

[8] Mester E, Szende B and Gartner P 1968 The effect of laser beams on the growth of hair in mice *Radiobiol. Radiother.* **9** 621–6

[9] Whelan H T *et al* 2001 Effect of NASA light-emitting diode irradiation on wound healing *J. Clin. Laser Med. Surg.* **19** 305–14

[10] Lubart R *et al* 1992 Effects of visible and near-infrared lasers on cell cultures *J. Photochem. Photobiol.* B **12** 305–10

[11] Gupta A, Dai T and Hamblin M R 2013 Effect of red and near-infrared wavelengths on low-level laser (light) therapy-induced healing of partial-thickness dermal abrasion in mice *Lasers Med. Sci.*

[12] Wu Q *et al* 2012 Low-level laser therapy for closed-head traumatic brain injury in mice: effect of different wavelengths *Lasers Surg. Med.* **44** 218–26

[13] Wong-Riley M T *et al* 2005 Photobiomodulation directly benefits primary neurons functionally inactivated by toxins: role of cytochrome c oxidase *J. Biol. Chem.* **280** 4761–71

[14] Karu T I 2010 Multiple roles of cytochrome c oxidase in mammalian cells under action of red and IR-A radiation *IUBMB Life* **62** 607–10

[15] Lane N 2006 Cell biology: power games *Nature* **443** 901–3

[16] Wang L, Zhang D and Schwarz W 2014 TRPV channels in mast cells as a target for low-level-laser therapy *Cells* **3** 662–73

[17] Wu Z H *et al* 2010 Mitochondrial signaling for histamine releases in laser-irradiated RBL-2H3 mast cells *Lasers Surg. Med.* **42** 503–9

[18] Nilius B *et al* 2007 Transient receptor potential cation channels in disease *Physiol. Rev.* **87** 165–217

[19] Chai B, Yoo H and Pollack G H 2009 Effect of radiant energy on near-surface water *J. Phys. Chem.* B **113** 13953–8

[20] Suda T, Takubo K and Semenza G L 2011 Metabolic regulation of hematopoietic stem cells in the hypoxic niche *Cell Stem Cell* **9** 298–310

[21] Spencer J A *et al* 2014 Direct measurement of local oxygen concentration in the bone marrow of live animals *Nature* **508** 269–73

[22] Mohle R *et al* 1999 Regulation of transendothelial migration of hematopoietic progenitor cells *Ann. NY Acad. Sci.* **872** 176–85 185–6

[23] Simsek T *et al* 2010 The distinct metabolic profile of hematopoietic stem cells reflects their location in a hypoxic niche *Cell Stem Cell* **7** 380–90

[24] Zhenwei J 2016 Mitochondria and pluripotent stem cells function *Yi Chuan* **38** 603–11

[25] Chen A C *et al* 2011 Low-level laser therapy activates NF-kB via generation of reactive oxygen species in mouse embryonic fibroblasts *PLoS One* **6** e 22453

[26] Owusu-Ansah E and Banerjee U 2009 Reactive oxygen species prime *Drosophila* haematopoietic progenitors for differentiation *Nature* **461** 537–41

[27] Min K H, Byun J H, Heo C Y, Kim E H, Choi H Y and Pak C S 2015 Effect of low-level laser therapy on human adipose-derived stem cells: in vitro and in vivo studies Aesth *Plast. Surg.* **39** 778–82

[28] Liao X, Xie G H, Liu H W, Cheng B, Li S H, Xie S, Xiao L L and Fu X B 2014 Helium-neon laser irradiation promotes the proliferation and migration of human epidermal stem cells in vitro: proposed mechanism for enhanced wound re-epithelialization *Laser Surg.* **32** 219–25

[29] O'Neill J 2015 Tackling a global health crisis: initial steps *The Review on Antimicrobial Resistance*

Photomedicine and Stem Cells
The Janus face of photodynamic therapy (PDT) to kill cancer stem cells, and photobiomodulation
(PBM) to stimulate normal stem cells
Heidi Abrahamse and Michael R Hamblin

Chapter 6

PBM and adipose-derived mesenchymal stem cells

6.1 Adipose-derived stem cells

Regenerative medicine and tissue engineering (TE) combine key elements such as biomaterials, stem cells, and bioactive agents (e.g. growth factors), in parallel with recent biotechnological advances. A constant and reliable source of autologous stem cells with pluripotent potential and ready availability will be required for these future cell-based therapies [1–4]. Bone marrow stem cells (BMSCs) have been extensively studied. but clinical application of these cells has presented problems including low cell number upon harvest, and pain and morbidity to the donor. Adipose tissue is derived from the mesenchyme and contains an easily isolated supportive stroma-containing stem cells, microvascular endothelial cells, and smooth muscle cells (SMCs) (table 6.1) [5, 6].

Mature adipocytes are easy to remove and separate from the stromal vascular fraction (SVF) by collagenase digestion and centrifugation, and the resulting cell population, termed ADSCs, are maintained in a non-inductive medium [3, 5, 8]. ADSCs are ideal for cellular therapy because they can be harvested, multiplied and handled easily, efficiently and non-invasively. ADSCs are pluripotent and proliferate considerably more than bone marrow (BM) mesenchymal cells, while morbidity to donors is less and a short wound healing time is required [8]. The most significant feature of adipose tissue is the relative expandability of this tissue that allows for large quantities of stem cells to be obtained [9]. ADSCs have multipotential differentiation capacity along the mesenchymal lineages of adipogenesis, osteogenesis, chondrogenesis, and myogenesis. Non-mesenchymal lineages have also been investigated and the transdifferentiation abilities of ADSCs confirmed that these cells can differentiate into bone, cartilage, fat, heart, nerve, liver, and smooth muscle (figure 6.1) [5–7, 9].

doi:10.1088/978-1-6817-4321-9ch6

Table 6.1. Comparison of the characteristics of ADSCs and BMSCs which have similar CD complements and differentiation potentials.

Surface marker	ADSCs	BMSCs	Reference
CD9	+	+	[5, 7]
CD10	+	+	
CD13	+	+	[5–7]
CD29	+	+	[5, 7]
CD31	−	−	[5–7]
CD34	−	−	
CD44	+	+	
CD45	−	−	
CD49d	+	−	[5, 7]
CD49e	+	+	
CD54	+	+	
CD55	+	+	
CD59	+	+	
CD90	+	+	[5–7]
CD105	+	+	
CD106	−	+	[5, 6]
CD117	+	+	[5]
CD146	+	+	[5, 7]
CD166	+	+	
STRO-1	+	+	[5, 6]
Differentiation potential			
Adipogenesis	+	+	[1, 5–7]
Osteogenesis	+	+	
Chondrogenesis	+	+	
Myogenesis	+	+	[1, 5, 7]
Cardiogenesis	+	+	
Neurogenesis	+	+	
Endothelial	+	+	
Hematopoietic	+	+	[7]

Cell therapies involving differentiation of SMCs may offer alternative treatment modalities for diseases that involve SMC pathology, such as gastrointestinal disease, urinary incontinence, cardiovascular disease, bladder dysfunction, hypertension, asthma, and many more [10, 11].

Adipose-derived stem cells (ADSCs) in regenerative medicine have been used for various cellular regeneration and TE studies. The nervous system has a limited capacity for self-repair, with mature nerve cells lacking the capacity to regenerate, and neural stem cells having a limited capability to generate new functional neurons in response to injury [12]. Wang *et al* [13] differentiated ADSCs into a neural cell lineage called olfactory ensheathing cells (OECs). They used a co-culture of ADSCs

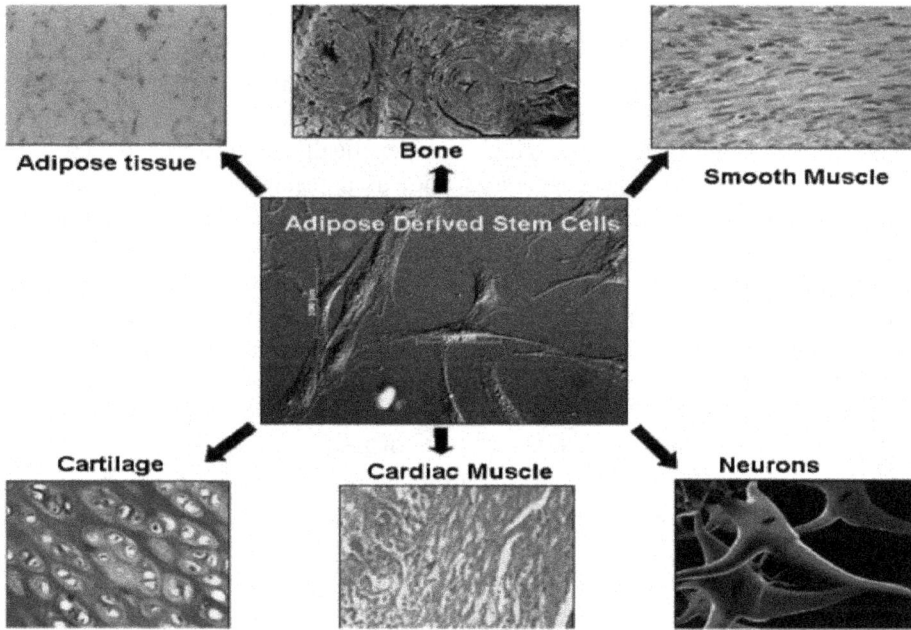

Figure 6.1. The multipotential differentiation capacity of ADSCs into bone, cartilage, fat, heart, nerve, and SMCs.

with OECs on three-dimensional scaffolds and observed that the differentiated cells had similar morphology and antigenic phenotypes (p75NTR+/ Nestin+/GFAP−) of OECs. Their results indicated that ADSCs had the potential to differentiate into OEC-like cells on three-dimensional scaffolds *in vitro*. OECs play a pivotal role in the repair of damaged central nervous system (CNS), thus the transplantation of OECs is a promising potential therapy for spinal cord injury. ADSCs are known to secrete multiple growth factors and therefore have cytoprotective effects in various injury models [4].

A study investigating the neuroprotective effects of ADSCs in a rat intracerebral hemorrhage (ICH) model indicated that ADSC transplantation promoted functional recovery, reduced apoptosis, and cerebral inflammation, and reduced brain atrophy and glial proliferation. The injected ADSCs had not differentiated into neuronal or glial lineages, but the majority differentiated into endothelial lineages leading rather to reduced chronic brain degradation and acute cerebral inflammation, and promoted long term functional recovery [4]. ADSCs also have the ability to survive, migrate and improve functional recovery after stroke in rats [12]. Neural differentiation was induced with azacytidine and exhibited expression of microtubule-associated protein 2 (MAP2) and glial fibrillary acidic protein (GFAP) and had neural morphology. In another study human ADSCs (hADSCs) labelled either with LacZ (β-galactosidase) or brain-derived neurotrophic factor (BDNF) adenoviruses injected into the lateral ventricle of the rat brain, showed that the transplanted cells migrated to different parts of the brain, there was an increase in migration to the injured cortex in ischemic brain injury by middle cerebral artery occlusion (MCAO), and there were improved functional deficits. Expression of MAP2 and GFAP in some of the transplanted cells was observed. BDNF-transduced hADSCs by intracerebral grafting drastically improved motor recovery of functional deficits in MCAO rats demonstrating that genetically engineered hADSCs can express biologically active gene products and, as a result, can function as useful vehicles for curative gene transfer to the brain.

Congenital and acquired heart disease has emerged as a leading cause of mortality and morbidity world-wide [5, 14]. Synthetic materials or bioprosthetic replacement devices for cardiovascular surgery are less than desirable since patients are subjected to various ongoing risks including thrombosis, re-operations, and limited durability due to a lack of growth in children and young adults. Cellular therapy using ADSCs is a potential new therapeutic option to treat cardiovascular disease [15] since they can differentiate into cardiomyocytes, hence these autologous adult stem cells (ASCs) are emerging as a new source of cells for cardiovascular repair [16]. Yamada et al [17] reported that CD29+ murine ADSCs could differentiate into cardiomyocytes with high efficiency based on characteristic morphology, electrophysical parameters, and molecular and protein expression. Transplanting CD29+ BAT-derived cells into the infarct border zone revealed that implanted cells expressed markers found in SMCs, endothelial cells, and cardiomyocytes by immunohistochemistry. Echocardiology showed improved ventricular function and reduced infarction area. Porcine BMSCs, as well as ADSCs transplanted into the infarct region, improved cardiac function and perfusion after intracoronary cell transplantation [18]. Myocardial infarction was induced followed by intracoronary injection of cultured autologous cells. This resulted in significantly decreased perfusion defects after 28 ± 3 days. A significant increase in left ventricular function, the thickness of the ventricular wall in the infarction area, and improved vascular density of the border zone were observed after the administration of stem cells. Co-localization of the grafted cells with von Willebrand factor and alpha smooth muscle actin was observed, with the incorporation into newly formed blood vessels.

Millions of people throughout the world are affected by bone diseases, such as osteoporosis and osteopenia. Repairing bone defects by gene and cell therapies and pharmacology warrants ongoing research [19]. A recent study by Elabd et al [19] indicated that multipotent human ADSCs can differentiate into osteocyte-like cells, that can develop into mineralized woven bone after four weeks when loaded on a hardening injectable bone substitute (HIBS) biomaterial and injected subcutaneously into nude mice. ADSCs were induced into oesteocytes in α-MEM (Eagle's minimum essential medium) containing a hormonal cocktail resulting in osteoblastic differentiation and expression of alkaline phosphatase, CBFA-1 (core binding factor alpha subunit 1), and osteonectin. Six-week-old nude mice were anaesthetized and injected with the HIBS, with or without ADSCs, and after 4 weeks the HIBS/cell implants exhibited a hard consistency compared to the cell-free/HIBS implants. All implants were fully colonized with woven bone within the ceramic in the intergranular spaces, and several cuboidal-shaped osteoblasts were present on the surface of the biphasic calcium phosphate particles associated with numerous osteocytes as revealed by histological analysis. Vessels that could support bone formation and multinucleated cells were extensively distributed. Duchenne muscular dystrophy (DMD) is an X-linked genetic disorder that is characterized by progressive muscle weakness and degeneration. Cell therapy is being pursued as a possible treatment modality for the repair of defective muscle in DMD [20]. Vieira *et al* [20] showed that ADSCs participate in myotube formation resulting in the restoration of dystrophin. Stem cells were grown in in appropriate media supplemented with dexamethasone, hydro-cortisone, and horse serum to differentiate hADSCs into the myogenic lineage. Immunofluorescence and RT-PCR studies showed expression of alpha-actinin in the differentiated culture, and revealed expression of myogenic determination (MyoD), telethonin, and dystrophin. Cocultures were also tested and results indicated that ADSCs participate in the generation of human myotubes through cellular fusion, since ADSCs plated with DMD myotubes revealed that ADSCs are able to fuse with DMD myotubes and restore dystrophin. ADSCs, when differentiated into muscle cells, can express dystrophin at a level equal to that of normal myoblasts.

There is a variety of diseases associated with SMC pathology. The use of stem cells for cell-based TE provides a promising possible alternate to current treatment strategies for smooth muscle repair or regeneration. A reliable source of SMCs for these applications remains a problem [10]. Rodriguez *et al* [10] differentiated processed lipoaspirate (PLA) cells into functional SMCs by culturing in smooth muscle induction media and heparin, which brought about genetic expression, at a transcriptional and translation level, of smooth muscle alpha actin (SMαa), SM22, caldesmon, calponin, smoothelin, and myosin heavy chain (MHC). In conclusion, their experiments indicated that ADSCs have the potential to differentiate into functional SMCs that could be used in therapeutic applications as a source of healthy SMCs. Kim *et al* [21] treated hADSCs with angiotensin II (Ang II), which increased the expression of smooth-muscle-specific genes, including SMα-actin, calponin, h-caldesmon, and SM-MHC, and also elicited the secretion of transforming growth factor-β1 (TGF-β1) and delayed phosphorylation of Smad2. Their work confirmed the importance of the MEK/ERK pathway in the Ang II-induced activation of the TGF-β1–Smad2 signaling pathway. These results suggest that Ang II induces differentiation of hADSCs into contractile smooth muscle-like cells through ERK-dependent activation of the autocrine TGF-β1–Smad2 crosstalk pathway.

The feasibility of hADSCs induced into SMCs *in vitro* as seeding cells in vascular TE have been investigated. ADSCs were subcultured, and platelet-derived growth factor (PDGF)-BB and TGF-β1 were added to enhance the SMC phenotype. Morphologically induced cells exhibited 'hill and valley' morphology, while the uninduced cells were similar to the passage zero (P0) ADSCs which had typical fibroblast-like morphology. SMC-specific markers including SMα-actin, SMMHC, and calponin were identified using immunohistochemistry and RT-PCR, indicating that hADSCs can be induced to express vascular smooth muscle markers, and they are a new potential source of vascular TE [22].

Although suppression of a patient's immune system is often performed in tissue and organ replacement surgeries, many other complications would arise. Thus, the aim of TE is to establish autologous tissue and organ transplants, i.e. from the patient's own cells. This will diminish graft versus host reactions [23]. ASCs such as ADSCs offer medical researchers and scientists the chance to produce autologous *de novo* tissues *ex vivo* with few ethical dilemmas. ADSCs isolated from a patient's own adipose tissue, cultured on with lineage-specific growth factors, and differentiated into a required tissue type, can be grafted back into the same patient [24]. Studies have shown that ADSCs display similar immunoregulatory properties to human BMSCs by inhibiting the proliferation and cytokine secretion of human primary T-cells in response to mitogens and allogenic T-cells [25]. ADSCs have two distinct advantages over previously used ASCs: ease of access and harvest, as well as improved proliferation potential [16].

Photobiomodulation (PBM) or low-intensity laser irradiation (LILI) can positively affect human ADSCs by increasing cellular proliferation, viability, and protein expression [26, 27]. These characteristics improves their potential in TE applications as the initial cell number could be increased before commencing differentiation leading to a higher yield of differentiated cells.

Smooth muscle is an active component of the cardiovascular, reproductive, urinary, and intestinal systems, and has been the subject of intense research in the field of regenerative medicine. Research has shown that SMCs that are obtained from diseased tissue can differ phenotypically and functionally from normal healthy SMCs, which consequently restricts their use [28, 29]. The findings of [10, 21, 22] describe a source of cells to use for SMC applications, as the results show that ADSCs have the potential to differentiate into functional SMCs and consequently may prove a useful source of autologous cells for reconstruction of diseased human organs and tissues containing smooth muscle [30].

PBM at different intensities has been shown to inhibit as well as stimulate cellular processes [31]. Studies on PBM and stem cells have shown that low-level lasers increase adenosine triphosphate (ATP) production, and migration [32]. PBM also promotes the proliferation of rat mesenchymal BM and cardiac stem cells *in vitro* [33] and can thus be used to stimulate the *in vitro* production of higher stem cell numbers. The addition of specific growth factors could enhance the differentiation of the stem cells into different cell types that could, in turn, be used in TE applications and reconstructive surgery. However, to be effective for use in TE, certain criteria need to be met, including that the cells of interest must be at high concentrations, easily harvested, and multipotent, while being able to differentiate into the required tissue and then be transplanted safely and effectively back into a host [34].

Epidermal growth factor (EGF) plays an important role in the regulation of cell growth, proliferation, and differentiation, yet is also involved in tumor proliferation, metastasis, apoptosis, angiogenesis, and wound healing [35], and participate in the development and differentiation of skin appendages, tissue repair, and modelling, and can activate ectodermal and mesodermal markers [36, 37]. Furthermore, EGF has been found to increase cellular proliferation and viability in precursor cells of the CNS [38], as well as being able to induce the phosphorylation of extracellular signal-regulated kinase pathways.

Studies on the effects of EGF on stem cells have found that pre-treatment of human ADSCs with EGF increased the differentiation potential of the cells along a neuronal lineage [39, 40] and in another stem cell model, it was shown that EGF stimulates the differentiation of hMSCs into bone-forming cells [41]. Little research has been focused on the effects of PBM in combination with EGF on ADSCs.

ADSCs remain in an undifferentiated state by suppression of intrinsic or extrinsic factors until stimulated to differentiate, and can be used for treatment of many diseases [42]. Diseases such as Parkinson's, stroke, multiple sclerosis, diabetes, and traumatic injuries are caused by either a loss of cells or damaged cells. ASCs could be used to treat these diseases and injuries, and since adipose tissue contains an abundant and accessible source of ASCs, and the stem cells can be isolated from this tissue, they become a valuable possible treatment regime [34, 43].

In a study by Stein et al [44], ADSCs were isolated from human adipose tissue, and the effects of PBM alone, as well as in combination with EGF, were evaluated. Western blot analysis confirmed that the isolated and cultured cells were stem cells by the expression of Thy-1, a stem cell marker known to be expressed by ADSCs. Thy-1 was expressed by the cells 24 h and 48 h after irradiation, whether cultured in the presence or absence of EGF. PBM was also found to promote proliferation and differentiation of human osteoblast cells in vitro at 632 nm with power output of 10 mW [44] as well as promote proliferation of mesenchymal and cardiac stem cells in culture at 1 J cm^{-2} and 3 J cm^{-2} [33]. Studies on low-level laser irradiation have shown that it could have a variety of photobiostimulatory effects, such as wound healing [45, 46], fibroblast proliferation [47–49], nerve regeneration [50], and collagen synthesis [51], and could increase migration of stem cells in vitro [32].

EGF as a growth factor regulates proliferation and differentiation [52], but is also involved in metastasis of cancer cells, programmed cell death, formation of blood vessels, wound healing, and tumor proliferation [35]. The addition of EGF to the cultures brought about an increase in cell viability and proliferation of which similar results were obtained by Svendsen et al [38], in which EGF-supplemented cultures showed an increase in the numbers of precursor cells of the CNS that remained viable for longer periods than in cultures lacking EGF. Hauner et al [53] found that incubating stromal cells from human adipose tissue in EGF-supplemented media had a potent stimulatory effect on cell proliferation; however, EGF also completely blocked the accumulation of lipids in these cells.

Additional studies showed the increase in cell proliferation was more pronounced in cells that had been cultured with 20 ng ml^{-1} of EGF and exposed to 5 J cm^{-2} PBM than in cells that had been cultured with the same concentration of EGF and not irradiated. The addition of EGF at different concentrations to other stem cell culture models has proven to be beneficial to the maintenance and proliferation of the cultured stem cells [54–56]. It would therefore be beneficial to the maintenance and expansion of ADSCs in culture if an optimal EGF concentration for this model could be determined; however, this requires further investigation.

6.2 PBM and ADSCs

PBM—a treatment modality that involves the application of low-power mono-chromatic and coherent light in the treatment of numerous diseases ranging from chronic musculoskeletal aches and dermatitis to the treatment and of chronic cutaneous wounds in diabetic patients—is a safe treatment that stimulate ASCs *in vivo* to proliferate and aid in the healing process. However, the long-term effects of the exposure of stem cells to PBM requires further investigation.

ADSCs are an abundant source of stem cells for TE and regenerative medicine, and the results presented here suggest that EGF in combination with PBM can increase the numbers and viability of cultured ADSCs. This is an important step in the expansion of stem cell numbers *in vitro*, especially in light of the potential role that ADSCs could play in regenerative medicine and TE, particularly in the use of autologous tissue transplants [57].

Stem cell treatment is becoming a promising therapy for many degenerative diseases [58]. One source of these cells is adipose tissue [59]. BMSCs were commonly used in scientific and clinical applications but, due to their limited number, differentiation potential limited by age [60], and invasive isolation procedure which may cause complications and death, ADSCs are now the preferred source [61]. ADSCs can be harvested from adipose tissue with ease and in abundance. These cells are easily cultured and maintain their mesenchymal stem cell pluripotency after many passages [6]. ADSCs are able to self-renew and differentiate into several lineages [62, 63]. Studies have shown that ADSCs could be differentiated into SMCs in the presence of the growth factors [10, 64]. These cells have also been differentiated into adipocytes, osteocytes, and neurons upon exposure to growth factors [65]. SMCs form smooth muscle tissues. These tissues are major components of systems like cardiovascular, reproductive, urinary, and intestinal systems. SMCs play a major role in diseases like cancer, asthma, arteriosclerosis, and hypertension since they constitute the main layer of smooth muscle tissues [10, 66]. Gastrointestinal smooth muscle diseases represent a major health problem affecting two million individuals every year [67]. Smooth muscle cell regeneration is required in the gastrointestinal tract as defects commonly occur [68]. PBM has shown different effects on several biological systems. It induces increased ATP production in mitochondria [69], elevation in collagen production in fibroblasts [70], and muscle regeneration processes following injury [71]. PBM has been shown to increase viability and proliferation of human fibroblast cells cultured in media with high glucose levels [72]. Cellular viability and proliferation have also been increased in ADSCs when exposed to PBM [73]. It has been shown to improve dental pulp stem cells when cultured in low nutritional conditions [74]. Growth factors are polypeptides that affect a number of cellular processes such as proliferation and differentiation both *in vivo* and *in vitro* [75]. Studies on retinoic acid have shown that it has several effects on cells including apoptosis, proliferation, differentiation, and maturation [76, 77]. Another growth factor, TGF-β1, plays a vital role in migration, angiogenesis, differentiation, proliferation, metastasis, and embryonic development [78, 79]. Beta1 integrin (CD29) is a protein that is encoded in humans by the ITGB1 gene [80]. It is associated with a late antigen receptor. It is expressed by ADSCs as a cell surface marker [81]. Thymocyte differentiation antigen 1 (Thy-1 CD90) is used as a marker for a variety of mesenchymal stem cells [82]. Both CD29 and CD90 are expressed by ADSCs and have been confirmed as mesenchymal stem cell markers

[83]. MHC is a cytoplasmic protein and a major component of SMCs [84]. It is a specific marker for smooth muscle differentiation. The expression of MHC is restricted to smooth muscle tissues [85]. Stem cells could be used for treatment of several diseases such as Parkinson's, stroke, diabetes, traumatic injury, and multiple sclerosis. These diseases are caused by either loss or damage of the cells in the organs or tissues [86]. Stem cells have to be differentiated into cells which are required to repair or replace the lost or damaged cells. ADSCs have been shown to have a high plasticity capability. They have been able to differentiate into smooth muscle, neuron, bone, cartilage, and fat cells [10, 87].

Differentiating ADSCs into SMCs would assist in the treatment of diseases that affect diseases in the cardiovascular, intestinal, urinary, and reproductive systems [66, 88]. Studies on coculturing of cells have proved that differentiation can be increased due to the secretion of growth factors of the cells that will be differentiated into. Previous studies have shown that PBM can increase cell viability and proliferation [73, 80]. The results of the present study showed a decrease in the proliferation of ADSCs and an increase in the proliferation of SMCs. This was observed through flow cytometry analysis; however, this could not be distinguished in the optical density analysis of the cocultures. In this study PBM increased the proliferation of cocultured cells; however, in the cocultures that had growth factors, proliferation decreased as compared to the cocultures without growth factors. This could have been due to the fact that proliferation of ADSCs was halted while differentiation into SMCs was initiated. Flow cytometry results concurred with this observation since a decrease in expression of the ADSCs markers in the cocultures was observed while those of the SMCs increased. This once again supports the argument that proliferation was inhibited since ADSCs were preparing for differentiation.

This study, in agreement with other studies done previously, found that PBM increases cell proliferation. PBM in combination with growth factors could differentiate ADSCs into SMCs. The study recommends that further investigations, in particular analyzing the regulation of different genes involved in the differentiation of ADSCs into SMCs, are necessary to confirm differentiation. Once differentiation is confirmed, PBM and growth factors, such as RA and TGF-β1, would play major roles in the established direct coculturing method for the differentiation of stem cells into SMCs and this would be very beneficial in the stem cell therapy for many degenerative diseases which involves smooth muscle cells. However, significant further research and investigation are required to realize the clinical potential for cell therapy of ADSC differentiation into SMCs and the contributory role that PBM may have in this process [89].

The augmentation of stem cell-based therapies to potentially modulate regenerative processes using non-invasive methods such as PBM holds great potential [90]. The development of stem cell therapy coincided with the development of other highly investigative and therapeutic disciplines such as tissue and genetic engineering, molecular biology and bio-compatible polymer synthesis leading to significant advances in the field of regenerative medicine. PBM has been applied clinically for the treatment of a variety of disorders including pain, inflammation, cancer, skin diseases, soft tissue injuries, and many more. It includes the use of low-power-intensity light devices delivering light in the visible to near-infrared wavelength of approximately 400–1000 nm. PBM of different intensities can both, inhibit or stimulate, cellular processes activating signaling cascades which ultimately lead to cellular modulation. Light energy is absorbed by light-absorbing molecules such as chromophores in the cells which direct and convert the light energy to be harvested in the form of chemical energy through the photochemical synthesis of ATP [91]. The mechanism whereby this occurs is not well understood but thought to be facilitated by the mitochondrial respiratory complexes resulting in production of reactive oxygen species (ROS), cyclic adenosine monophosphate (cAMP) synthesis and influx of intracellular calcium. A significant increase in ATP production has been identified using a range of different wavelengths including 632. 8, 830, and 904 nm, and in PBM of a number of different cell types such as fibroblasts, keratinocytes, osteoblasts, lymphocytes, and endothelial cells [91, 92].

Peptide-based biopolymers and the application of these engineered biomolecules include TE as they serve as injectable scaffolds that form gels *in vivo* via physical or chemical means, and provide a minimally invasive route to deliver tissue scaffolds [93]. Hollow, bioresorbable, bioactive, ceramic tricalcium phosphate (TCP) shells impregnated with a bioresorbable polymer, polylactide-coglycolide (PLGA), was used to investigate the potential as bulking and carrier particles for stem cell therapy.

6.3 Studies from the Abrahamse laboratory

Work from the Abrahamse laboratory has focused on the ability of laser irradiation to proliferate and maintain ADSC character and increase the rate of proliferation and maintenance of differentiation of ADSCs into SMCs. In addition, wavelength and fluence have also been studied and found to contribute to biostimulation of ADSCs, although wavelengths of higher than 636 nm and greater fluencies including 10 and 15 J cm^{-2} decreases viability and proliferation. The use and application of ADSC differentiated SMCs for clinical applications using resorbable, injectable solid PCL microspheres as a delivery and bulking agent for clinical applications shows great promise and has been shown to have potential in other applications including bone disease, where human mesenchymal stem cells were cultured on PCL–TCP scaffolds, and cell attachment, spreading, and cellular bridging have been observed [94]. Growth factors may be incorporated into the polymer matrix for controlled release at the implanted site. The purpose of the growth factors is to promote regeneration of tissue in the affected area, leading to long-term bulking. The polymer properties and manufacturing parameters can be selected to manipulate the bioresorption rate and growth factor release rate. The matrix also provides the opportunity for the incorporation and controlled release of therapeutic drugs at the target site [95]. In conclusion, the enormous potential that ADSCs hold due to their specific characteristics in conjunction with PBM to further enhance and increase their ability to improve TE and regenerative medicine is demonstrated in figure 6.2. It shows the isolation of ADSCs from a patient, the isolation, characterization, and differentiation into the diseased tissue required by the patient, and the

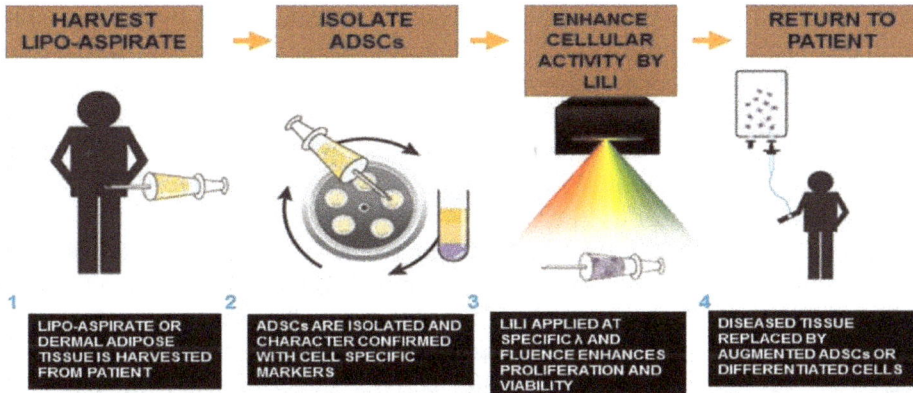

Figure 6.2. Autologous stem cell therapy augmented by PBM.

enhancement that PBM offers in the process of improved viability, proliferation, and differentiation prior to transplantation back into the patient by autologous grafts [96].

6.4 Work from the Hamblin laboratory

Work from the Hamblin laboratory has focused on *in vitro* studies of ADSCs and particularly on the differentiation into cells of the osteogenic lineage. The first study [97] looked at the effects of four different wavelengths of light (420 nm, 540 nm, 660 nm, 810 nm) at the same dose (3 J cm^{-2}) delivered five times (every two days) on the differentiation of human ADCSs. Cells were cultured in osteogenic medium over three weeks. The expression of the following osteogenic transcription factors was measured by RT-PCR: RUNX2, osterix, and the osteoblast protein, osteocalcin. The 420 nm and 540 nm wavelengths (blue and green) were more effective in stimulating osteoblast differentiation compared to 660 nm and 810 nm (red and NIR). Intracellular calcium was higher after 420 nm and 540 nm, and could be inhibited by pharmacological inhibitors of transient receptor potential (TRP) ion channels. Capsazepine is an inhibitor of TRPV1 (TRP vanilloid type) while SKF96365 is a broad-spectrum inhibitor of ion channels. These inhibitors also inhibited osteogenic differentiation, suggesting that TRP ion channels are involved in ADSC osteogenic differentiation, and that these light-gated calcium ion channels could be activated by PBM using blue and green wavelengths.

The next study [98] compared the effects of two different NIR wavelengths (810 nm and 980 nm) on hADSCs *in vitro*. Both wavelengths showed a biphasic dose response for stimulation of proliferation at 24 h, but 810 nm had a peak dose response at 3 J cm^{-2}, while the peak dose for 980 nm was 10–100 times lower at 0.03 or 0.3 J cm^{-2}. Moreover, 980 nm (but not 810 nm) increased cytosolic calcium while decreasing mitochondrial calcium. 810 nm gave a larger increase in mitochondrial membrane potential and ATP production. Both wavelengths produced an increase in ROS, but the increase with 980 nm was faster. The effects of 980 nm could be blocked by calcium channel blockers (capsazepine for TRPV1 and SKF96365 for

TRPC channels), which had no effect on proliferation caused by 810 nm. To test the hypothesis that the chromophore for 980 nm was intracellular water, which could possibly form a microscopic temperature gradient upon laser irradiation, we added cold medium (4 °C) during the light exposure, or pre-incubated the cells at 42 °C, both of which abrogated the effect of 980 nm but not 810 nm. We conclude that 980 nm affects temperature-gated calcium ion channels, while 810 nm largely affects mitochondrial cytochrome c oxidase.

The final study looked at the effects of four visible wavelengths of light (420 nm, 540 nm, 660 nm, 810 nm) on the proliferation of hADSCs [99]. When cultured in proliferation medium there was a clear difference between blue/green which inhibited proliferation and red/NIR which stimulated proliferation, all at 3J cm^{-2}. Blue/green reduced cellular ATP, while red/NIR increased ATP in a biphasic manner. Blue/green produced a bigger increase in intracellular calcium and ROS. Blue/green reduced mitochondrial membrane potential and lowered intracellular pH, while red/NIR had the opposite effect. The TRPV1 ion channel was expressed in hADSC, and the TRPV1 ligand capsaicin (5uM) stimulated proliferation, which could be abrogated by capsazepine. The inhibition of proliferation caused by blue/green could also be abrogated by capsazepine, and by the antioxidant, N-acetylcysteine. The data suggest that blue/green light inhibits proliferation by activating TRPV1, and increasing calcium and ROS.

References

[1] Gimble J M, Katz A J and Bunnell B A 2007 Adipose-derived stem cells for regenerative medicine *Circulation Res.* **100** 1249–60

[2] Sandor G K B and Suuronen R 2008 Combining adipose derived stem cells, resorbable scaffolds and growth factors: an overview of tissue engineering *JCDA* **74** 167–70

[3] Zuk P A, Zhu M, Mizuno H, Huang J, Futrell J W, Katz A J, Benhaim P, Lorenz H P and Hedrick M H 2001 Multilineage cells from human adipose tissue: implications for cell-based therapies *Tissue Eng.* **7** 211–28

[4] Kim J M *et al* 2007 Systemic transplantation of human adipose stem cells attenuated cerebral inflammation and degeneration in a hemorrhagic stroke model *Brain Res.* **1183** 43–50

[5] Strem B M, Hicok K C, Zhu M, Wulur I, Alfonso Z, Schreiber R E, Fraser J K and Hedrick M H 2005 Multipotential differentiation of adipose tissue-derived stem cells *Keio J. Med.* **54** 132–41

[6] Zuk P A, Zhu M, Ashjian P, De Ugarte D A, Huang J I, Mizuno H, Alfonso Z C, Fraser J K, Benhaim P and Hedrick M H 2002 Human adipose tissue is a source of multipotent stem cells *Mol. Biol. Cell* **13** 4279–95

[7] Schäffler A and Büchler C 2008 Concise review: adipose tissue-derived stromal cells—basic and clinical implications for novel cell-based therapies *Stem Cells* **25** 818–27

[8] Ogawa R 2006 The importance of adipose-derived stem cells and vascularised tissue regeneration in the field of tissue transplantation *Curr. Stem Cell Res. Ther.* **1** 13–20

[9] Fraser J K, Wulur I, Alfonso Z and Hedrick M H 2006 Fat tissue: an underappreciated source of stem cells for biotechnology *Trends Biotechnol.* **24** 150–4

[10] Rodríguez L V, Alfonso Z, Zhang R, Leung J, Wu B and Ignarro L J 2006 Clonogenic multipotent stem cells in human adipose tissue differentiate into functional smooth muscle cells *Proc. Natl. Acad. Sci. USA* **103** 12167–72

[11] Wong J Z, Woodcock-Mitchell J, Mitchell J, Rippetoe P, White S, Absher M, Baldor L, Evans J, McHugh K M and Low R B 1998 smooth muscle actin and myosin expression in cultured airway smooth muscle cells *Am. J. Physiol.* **274** L786–92

[12] Kang S K, Lee D H, Bae Y C, Kim H K, Baik S Y and Jung J S 2003 Improvement of neurological deficits by intracerebral transplantation of human adipose tissue-derived stromal cells after cerebral ischemia in rats *Exp. Neurol.* **183** 355–66

[13] Wang B, Han J, Gao Y, Xiao Z, Chen B, Wang X, Zhao W and Dai J 2007 The differentiation of rat adipose-derived stem cells into OEC-like cells on collagen scaffolds by co-culturing with OECs *Neurosci. Lett.* **421** 191–6

[14] Wu K, Liu Y L, Cui B and Han Z 2006 Application of stem cells for cardiovascular grafts tissue engineering *Transpl. Immunol.* **16** 1–7

[15] Bai X, Pinkernell K, Song Y H, Nabzdyk C, Reiser J and Alt E 2006 Genetically selected stem cells from human adipose tissue express cardiac markers *Biochem. Biophys. Res. Commun.* **353** 665–71

[16] Sanz-Ruiz R, Santos M E, Muñoa M D, Martín I L, Parma R, Fernández P L and Fernández-Avilés F 2008 Adipose tissue-derived stem cells: the friendly side of a classic cardiovascular foe *J. Cardiovasc. Transl. Res.* **1** 55–63

[17] Yamada Y, Wang X D, Yokoyama S, Fukuda N and Takakura N 2006 Cardiac Progenitor cells in brown adipose tissue repaired damaged myocardium *Biochem. Biophys. Res. Commun.* **342** 662–70

[18] Valina C, Pinkernell K, Song Y H, Bai X, Sadat S, Campeau R J, Le Jemtel T H and Alt E 2007 Intracoronary administration of autologous adipose tissue-derived stem cells improves left ventricular function, perfusion, and remodelling after acute myocardial infarction *Eur. Heart J.* **28** 2667–77

[19] Elabd C *et al* 2007 Human adipose tissue-derived multipotent stem cells differentiate *in vitro* and *in vivo* into osteocyte-like cells *Biochem. Biophys. Res. Commun.* **361** 342–8

[20] Vieira N M, Brandalise V, Zucconi E, Jazedje T, Secco M, Nunes V A, Strauss B E, Vainzof M and Zatz M 2008 Human multipotent adipose derived stem cells restore dystrophin expression of Duchenne skeletal muscle cells *in vitro Biol. Cell* **100** 231–41

[21] Kim Y M, Jeon E S, Kim M R, Jho S K, Ryu S W and Kim J H 2008 Angiotensin II-induced differentiation of adipose tissue-derived mesenchymal stem cells to smooth muscle like cells *Int. J. Biochem. Cell. Biol.* **40** 2482–91

[22] Yang P, Yin S, Cui L, Li H, Wu Y, Liu W and Cao Y 2008 Experiment of adipose derived stem cells induced into smooth muscle cells *Zhongguo Xiu Fu Chong Jian Wai Ke Za Zhi* **22** 481–6

[23] Ringe J, Kaps C, Burmester G R and Sittinger M 2002 Stem cells for regenerative medicine: advances in the engineering of tissues and organs *Naturwissenschaften* **89** 338–51

[24] Moore T J 2007 Stem cell Q and A—an introduction to stem cells and their role in scientific and medical research *Med. Technol. SA* **21** 3–6

[25] Yanez *et al* 2008

[26] Mvula B, Mathope T, Moore T and Abrahamse H 2007 The effect of low level laser therapy on adipose derived stem cells *Lasers Med. Sci.* **23** 277–82

[27] Mvula B, Moore T J and Abrahamse H 2009 Effect of low-level laser irradiation and epidermal growth factor on adult human adipose-derived stem cells *Lasers Med. Sci.* **25** 33–9

[28] Dozmorov M G, Kropp B P, Hurst R E, Cheng E Y and Lin H K 2007 Differentially expressed gene networks in cultured smooth muscle cells from normal and neuropathic bladder *J. Smooth Muscle Res.* **43** 55–72

[29] Lin H K, Cowan R, Moore P, Zhang Y, Yang Q, Peterson J A Jr, Tomasek J J, Kropp B P and Cheng E 2004 Characterization of neuropathic bladder smooth muscle cells in culture *J. Urol.* **171** 1348–52

[30] De Villiers J, Houreld N and Abrahamse H 2009 Adipose derived stem cells and smooth muscle cells: implications for regenerative medicine *Stem Cell Rev. Rep.* **5** 256–65

[31] Moore P, Ridgway T D, Higbee R G, Howard E W and Lucroy M D 2005 Effect of wavelength on low-intensity laser irradiation stimulated cell proliferation in vitro *Lasers Surg. Med.* **36** 8–12

[32] Gasparyan L, Brill G and Makela A 2004 Influence of low level laser radiation on migration of stem cells *Proc. SPIE* **5968** 58–63

[33] Tuby H, Maltz L and Oron U 2007 Low level laser irradiation (LLLI) promotes proliferation of mesenchymal and cardiac stem cells in culture *Lasers Surg. Med.* **39** 373–8

[34] Gimble J M and Guilak F 2003 Adipose derived adult stem cells: isolation, characterisation, and differentiation potential *Cytotherapy* **5** 362–9

[35] Bouis D, Kusumanto Y K, Meijer C, Mulder N H and Hospers G A P 2006 A review on pro- and anti-angiogenic factors as targets of clinical intervention *Pharmacol. Res.* **53** 89–103

[36] Li O, Vasudevan D, Davey C A and Droge P 2006 High-level expression of DNA architectural factor HMGA2 and its association with nucleosomes in human embryonic stem cells *Genesis* **44** 523–9

[37] Fu X B, Sun X Q, Sun T Z, Dong Y H, Gu X M, Chen W and Sheng Z Y 2002 Epidermal growth factor stimulates tissue repair in skin through skin stem cell activation Zhongguo Xiu Fu Chong Jian Wai ke Za Zhi **6** 31–5

[38] Svendsen C N, Fawcett J W, Bentlage C and Dunnett S B 1995 increased survival of rat EGF-generated CNS precursor cells using B27 supplemented medium *Exp. Brain Res.* **102** 407–14

[39] Safford K M, Hicok K C, Safford S D, Halvorsen Y D, Wilkison W O, Gimble J M and Rice H E 2002 Neurogenic differentiation of murine and human adipose derived stromal cells *Biochem. Biophys. Res. Commun.* **294** 371–9

[40] Angenieux B, Schrorderet D F and Arsenijevic Y 2006 Epidermal growth factor is a neuronal differentiation for retinal stem cells *in vitro Stem Cells* **24** 696–706

[41] Kratchmarova I, Blagoev B, Haack-Sorensen M, Kassem M and Mann M 2005 Mechanisms of divergent growth factor effects in mesenchymal stem cell differentiation *Science* **308** 1472–7

[42] Rao M S 1999 Multipotent and restricted precursors in the central nervous system *Anat. Rec.* **257** 137–48

[43] Safford K M and Rice H E 2005 Stem cell therapy for neurologic disorders: therapeutic potential of adipose-derived stem cells *Curr. Drug Targets* **6** 57–62

[44] Stein A, Benayahu D, Maltz L and Oron U 2005 Low level laser irradiation promotes proliferation and differentiation of human osteoblasts *in vitro Photomed. Laser Surg.* **23** 161–6

[45] Houreld N and Abrahamse H 2005 Low level laser therapy for diabetic foot wound healing *Diabet. Foot J* **8** 182–93

[46] Hawkins D, Houreld N and Abrahamse H 2005 Low level laser therapy (LLLT) as an effective therapeutic modality for delayedwound healing *Ann. NY Acad. Sci.* **1056** 486–93

[47] Kana J S, Hutschenreiter G, Haina D and Waidelich W 1981 effect of low power density laser radiation on healing on open skin wound in rats *Arch. Surg.* **116** 293–6

[48] Boulton M and Marshall J 1986 He–Ne laser stimulation of human fibroblast proliferation and attachment *in vitro Lasers Life Sci.* **1** 123–34

[49] Van Breugel H and Bar P R D 1992 Power density and exposure time of He–Ne laser irradiation are more important than total energy dose in photobiomodulation of human fibroblasts *in vitro Lasers Surg. Med.* **12** 528–37

[50] Anders J J, Borke R C, Woolery S K and Van der Merwe W P 1993 Low power laser irradiation alters the rate of regeneration of rat facial nerve *Lasers Surg. Med.* **13** 72–82

[51] Lam T S, Abergel R P, Meeker C A, Castel J C, Ewyer R M and Uitto J 1986 Laser stimulation of collagen synthesis in human skin fibroblasts culture *Lasers Life Sci.* **1** 61–77

[52] Carpenter G and Cohen S 1990 Epidermal growth factor *J. Biol. Chem.* **265** 7709–12

[53] Hauner H, Rohrig K and Petruschke T 1995 Effects of epidermal growth factor (EGF), platelet-derived growth factor (PDGF) and fibroblast growth factor (FGF) on human adipocyte development and function *Eur. J. Clin. Invest.* **25** 90–6

[54] Pitman M, Emery B, Binder M, Wang S, Butzkueven H and Kilpatrick T J 2004 LIF receptor signaling modulates neural stem cell renewal *Mol. Cell. Neurosci.* **27** 255–66

[55] Tropepe V, Sibilia M, Ciruna B G, Rossant J, Wagner E F and van der Kooy D 1999 Distinct neural stem cells proliferate in response to EGF and FGF in the developing mouse telencephalon *Dev. Biol.* **208** 166–88

[56] Heo J S, Lee J L and Han J H 2005 EGF stimulates proliferation of mouse embryonic stem cells: involvement of calcium ions influx and P44/42 MAPKS *Am. J. Physiol. Cell. Physiol.* **290** C123–C133

[57] Mvula B, Moore T J and Abrahamse H 2010 Effect of low-level laser irradiation and epidermal growth factor on adult human adipose-derived stem cells *Lasers Med. Sci.* **25** 33–9

[58] Taha M F 2010 Cell based-gene delivery approaches for the treatment of spinal cord injury and neurodegenerative disorders *Curr. Stem Cell Res. Ther.* **5** 23–36

[59] Nakagami H, Morishita R, Maeda K, Kikuchi Y, Ogihara T and Kaneda Y 2006 Adipose tissue-derived stromal cells as a novel option for regenerative cell therapy *J. Atheroscler. Thromb.* **13** 77–81

[60] Mueller S M and Glowacki J 2001 Age-related decline in the osteogenic potential of human bone marrow cells cultured in three-dimensional collagen sponges *J. Cell. Biochem.* **82** 583–90

[61] Lee S Y, Miwa M, Sakai Y, Kuroda R, Matsumoto T, Iwakura T, Fujioka H, Doita M and Kurosaka M 2007 *In vitro* multipotentiality and characterization of human unfractured traumatic hemarthrosis-derived progenitor cells: a potential cell source for tissue repair *J. Cell. Physiol.* **210** 561–6

[62] Jang S, Cho H H, Cho Y B, Park J S and Jeong H S 2010 Functional neural differentiation of human adipose tissue derived stem cells using bFGF and forskolin *BMC Cell Biol.* **11** 25

[63] Huang T, He D, Kleiner G and Kuluz J 2007 Neuron-like differentiation of adipose-derived stem cells from infant piglets in vitro *J. Spinal Cord Med.* **30** 35–40

[64] De Villiers J, Houreld N and Abrahamse H 2011 Influence of low intensity laser irradiation on isolated human adipose derived stem cells over 72 hrs and their differentiation potential into smooth muscle cells using retinoic acid *Stem Cell Rev. Rep.* **7** 869–82

[65] Hu L, Hu J, Zhao J, Liu J, Ouyang W, Yang C, Gong N, Du L, Khanal A and Chen L 2013 Side-by-side comparison of the biological characteristics of human umbilical cord and adipose tissue-derived mesenchymal stem cells *Biomed. Res. Int.* **2013** 438243

[66] Sinha S, Wamhoff B R, Hoofnagle M H, Thomas J, Neppl R L, Deering T, Helmke B P, Bowles D K, Somlyo A V and Owens G K 2006 Assessment of contractility of purified smooth muscle cells derived from embryonic stem cells *Stem Cells* **24** 1678–88

[67] Gabetta V, Trzyna W, Phiel C and McHugh K M 2003 Vesicle associated protein-a is differentially expressed during intestinal smooth muscle cell differentiation *Dev. Dyn.* **228** 11–20

[68] McHugh K M 1996 Molecular analysis of gastrointestinal smooth muscle development *J. Pediatr. Gastroenterol. Nutr.* **23** 379–94

[69] Passarella S, Casamassima E, Molinari S, Pastore D, Quagliariello E, Catalano I M and Cingolani A 1984 Increase of proton electrochemical potential and ATP synthesis in rat liver mitochondria irradiated *in vitro* by helium-neon laser *FEBS Lett.* **175** 95–9

[70] Kovács I B, Mester E and Görög P 1974 Stimulation of wound healing with laser beam in the rat *Experientia* **30** 1275–6

[71] Amaral A C1, Parizotto N A and Salvini T F 2001 Dose dependency of low-energy HeNe laser effect in regeneration of skeletal muscle in mice *Lasers Med. Sci.* **16** 44–51

[72] Esmaeelinejad M1, Bayat M, Darbandi H, Bayat M, Mosaffa N 2014 The effects of low-level laser irradiation on cellular viability and proliferation of human skin fibroblasts cultured in high glucose mediums *Lasers Med. Sci.* **29** 121–9

[73] Mvula B, Mathope T, Moore T and Abrahamse H 2008 The effect of low level laser irradiation on adult human adipose derived stem cells *Lasers Med. Sci.* **23** 277–82

[74] Eduardo Fde P, Bueno D F, de Freitas P M, Marques M M, Passos-Bueno M R, Eduardo Cde P and Zatz M 2008 Stem cell proliferation under low intensity laser irradiation: a preliminary study *Lasers Surg. Med.* **40** 433–8

[75] Goustin A S, Leof E B, Shipley G D and Moses H L 1986 Growth factors and cancer *Cancer Res.* **46** 1015–29

[76] Duong V and Rochette-Egly C 2011 The molecular physiology of nuclear retinoic acid receptors. From health to disease *Biochim. Biophys. Acta.* **1812** 1023–31

[77] Gudas L J and Wagner J A 2011 Retinoids regulate stem cell differentiation *J. Cell. Physiol.* **226** 322–30

[78] Zhang R, Jack G S, Rao N, Zuk P, Ignarro L J, Wu B and Rodríguez L V 2012 Nuclear fusion-independent smooth muscle differentiation of human adipose-derived stem cells induced by a smooth muscle environment *Stem Cells* **30** 481–90

[79] Rahimi R A and Leof E B 2007 TGF-β signaling: a tale of two responses *J. Cell. Biochem.* **102** 593–608

[80] Goodfellow P J, Nevanlinna H A, Gorman P, Sheer D, Lam G and Goodfellow P N 1989 Assignment of the gene encoding the beta-subunit of the human fibronectin receptor (β-FNR) to chromosome 10p11.2 *Ann. Hum. Genet.* **53** 15–22

[81] Yoshimura K *et al* 2006 Characterization of freshly isolated and cultured cells derived from the fatty and fluid portions of liposuction aspirates *J. Cell. Physiol.* **208** 64–76

[82] Masson N M, Currie I S, Terrace J D, Garden O J, Parks R W and Ross J A 2006 Hepatic progenitor cells in human fetal liver express the oval cell marker Thy-1 *Am. J. Physiol. Gastrointest. Liver Physiol.* **291** G45–G54

[83] Zahran F, Salam M, Loffy A and El-Deen I M 2012 isolation and characterisation of adipose tissue-derived stem cells: an *in vitro* study *Basic Res. J. Med. Clin. Sci.* **1** 88–94

[84] Nicolas M M, Tamboli P, Gomez J A and Czerniak B A 2010 Pleomorphic and dedifferentiated leiomyosarcoma: clinicopathologic and immunohistochemical study of 41 cases *Hum. Pathol.* **41** 663–71

[85] Aikawa M, Sakomura Y and Ueda M 1997 Re-differentiation of smooth muscle cells after coronary angioplasty determined via myosin heavy chain expression *Circulation* **96** 82–90

[86] Safford K M and Rice H E 2005 Stem cell therapy for neurologic disorders: therapeutic potential of adipose-derived stem cells *Curr. Drug Targ.* **6** 57–62

[87] Serakinci N and Keith W N 2006 Therapeutic potential of adult stem cells *Eur. J. Cancer* **42** 1243–6

[88] Carmeliet P 2000 Mechanisms of angiogenesis and arteriogenesis *Nat. Med.* **6** 389–95

[89] Mvula B and Abrahamse H 2014 Low intensity laser irradiation and growth factors influence differentiation of adipose derived stem cells into smooth muscle cells in a coculture environment over a period of 72 hours *Int. J. Photoenergy* **2014** 598793

[90] Lin F *et al* 2010 Lasers, stem cells, and COPD *J. Translat. Med.* **8** 16–26

[91] Gao X and Xing D 2009 Molecular mechanisms of cell proliferation induced by low power laser irradiation *J. Biomed. Sci.* **16** 4–30

[92] Drochioiu G 2010 Laser induced ATP formation: mechanism and consequences *Photomed. Laser Surg.* **28** 573–4

[93] Chow D, Nunalee M L, Lim D W, Simnick A J and Chilkoti A 2008 Peptide-based biopolymers in biomedicine and biotechnology *Mater. Sci. Eng.* R **62** 125–55

[94] Rai B, Lin J L, Lim Z X, Guldberg R E, Hutmacher D W and Cool S M 2010 Differences between *in vitro* viability and differentiation and *in vivo* bone-forming efficacy of human mesenchymal stem cells cultured on PCL-TCP scaffolds *Biomaterials* **31** 7960–70

[95] Abrahamse H 2010 Low intensity laser irradiation ameliorates stem cell based therapy for use in autologous grafts *ScienceMED* **1** 1–6

[96] Abrahamse H, Houreld N N, Muller S and Ndlovu L 2010 Fluence and wavelength of low intensity laser irradiation affect activity and proliferation of human adipose derived stem cells *Med. Technol. SA* **24** 8–14

[97] Wang Y *et al* 2016 Photobiomodulation (blue and green light) encourages osteoblastic-differentiation of human adipose-derived stem cells: role of intracellular calcium and light-gated ion channels *Sci. Rep.* **6** 33719

[98] Wang Y *et al* 2016 Photobiomodulation of human adipose-derived stem cells using 810 nm and 980 nm lasers operates via different mechanisms of action *Biochim. Biophys. Acta* **1861** 441–9

[99] Wang Y *et al* 2017 Red (660 nm) or near-infrared (810 nm) photobiomodulation stimulates, while blue (415nm), green (540 nm) light inhibits proliferation in human adipose-derived stem cells *Sci. Rep.* **7** in press

Photomedicine and Stem Cells

The Janus face of photodynamic therapy (PDT) to kill cancer stem cells, and photobiomodulation
(PBM) to stimulate normal stem cells

Heidi Abrahamse and Michael R Hamblin

Chapter 7

PBM and dental stem cells

7.1 Introduction

One of the long-held dreams in the field of dentistry has been the ability to grow new teeth in adults. One way of accomplishing this goal may be to activate endogenous dental stem cells in a minimally invasive manner [1–3]. Figure 7.1 illustrates how photobiomodulation (PBM) may work in concert with dental stem cells and growth factors in order to reconstruct either entire teeth, dental pulp, or periapical tissues.

When PBM is delivered to biological cells and tissue, photon absorption leads to generation of transient, extremely reactive chemical intermediates, termed reactive oxygen species (ROS) [4]. The mitochondrial enzyme, cytochrome c oxidase (CCO), is among the best characterized biological photo-acceptors absorbing strongly in the red (e.g. 632 nm) and in the near-infrared (NIR), and has key functions in the mitochondrial electron transport chain modulating ROS and adenosine triphosphate (ATP) synthesis [5]. When the mitochondrial membrane potential is increased above the baseline due to stimulation of CCO activity, a relatively brief burst of ROS is produced [6]. On the other hand, when the mitochondrial membrane potential drops below the baseline, as frequently occurs in various pathological states, then ROS are produced on a chronic long-term basis and this condition is termed 'oxidative stress'. When CCO absorbs PBM light and the mitochondrial membrane is restored towards normal baseline levels, then the production of ROS falls rather than rises [7].

Members of the transforming growth factor-β (TGF-β) family have been suggested to mediate some biological effects of PBM [8–12]. The TGF-β family consists of over 75 members reported to date encoded by 23 distinct genes including the bone morphogenetic proteins (BMPs), TGF-βs, activin, nodal, inhibin, and growth/differentiation factors (GDFs) among others [13–15]. The interactions and complex context-dependent roles of these factors in maintaining stem cell pluripotency as well as inducing differentiation has been elegantly demonstrated using a

doi:10.1088/978-1-6817-4321-9ch7

(a). Regeneration of Entire Tooth

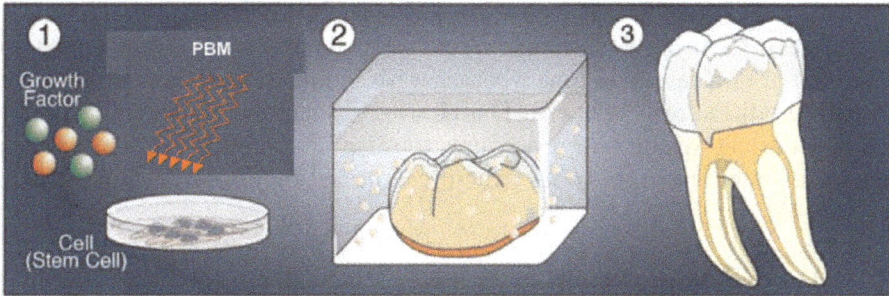

(b). Regeneration of Dental Pulp

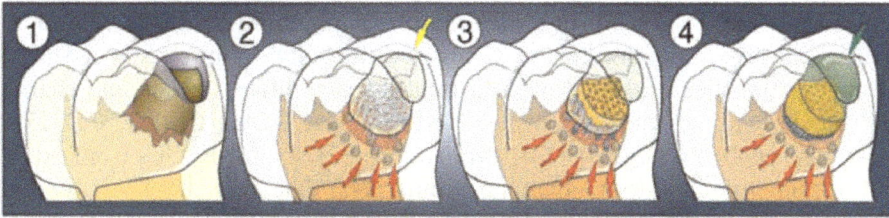

(c). Regeneration of Periapical Tissues

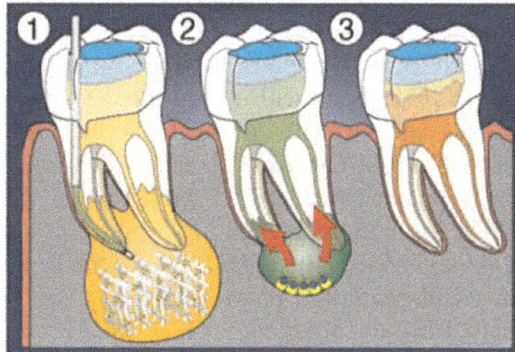

Figure 7.1. Strategies for the regeneration of dental pulp and periapical tissues using the combination of stem cells, growth factors, and PBM. (a) Regeneration of the entire tooth. In the strategy, a tooth germ is regenerated by growth factors, and stem cells in organ culture. (b) Regeneration of dental pulp and dentin. This strategy is further classified into two ways. One is the regeneration by the combination of supplied tissue engineering (TE) factors including growth factors, scaffolds, and stem cells. The other is the regeneration by supplied growth factors and scaffolds, and stem or progenitor cells are induced from residual tissues. (c) Regeneration of periapical tissues by the combination of TE factors with external stimulation such as PBM using laser irradiation.

variety of experimental models [16–18]. It is clear that specific mediators can act in a domino-like manner in developmental processes, initiating a cascade of downstream biological effects [19, 20], and recent studies have specifically highlighted the role of the TGF-β signaling pathway as a master regulator of stem cell differentiation [21, 22]. TGF-β family members play a central role in tooth development, and specifically in pulp–dentin homeostasis [23, 24]. In addition to their role in the homeostasis of normal tooth structure, adult stem cells in the tooth pulp and periodontal ligaments are capable of multi-lineage differentiation into cells that can form bone, cartilage, or adipose tissue [25–30]. Dentin repair is a critical factor that can determine whether teeth should be surgically removed, or if they can be satisfactorily restored. This choice between removal of teeth and fitting of prosthetics and restorative procedures to maintain viability and normal neuro-mechanical functions of the tooth structure underlies millions of dental procedures each year [31].

7.2 PBM and dental stem cells

A multipotent niche for human dental stems cells (hDSCs) resides within the tooth pulp. These stem cells contribute to the normal pathophysiological role of tooth pulp, specifically in dentin repair and regeneration. hDSCs possess broad pluripotent potential as demonstrated by their multi-lineage differentiation and characteristic cell surface marker profiles (positive for CD44, CD90, CD106, CD117, and Stro-1, while CD45 is negative) [25]. Arany *et al* [32] isolated hDSCs from extracted tooth specimens and assessed their surface markers that were characteristic of their pluripotent stem cell state over multiple passages. These hDSC cells were used in differentiation experiments between 3–7 passages as their potency and stability has been shown to vary over passages [33]. TGF-β1 binding to its receptor leads to phosphorylation of smad to produce phospho-smad. hDSC cells were responsive to PBM irradiation and TGF-β1 treatments as indicated by phospho-smad activation and translocation. Pretreatment with antioxidants such as N-acetyl-cysteine (a compound which quenches ROS) abrogated smad activation. The same abrogation of smad activation was obtained when a TGF-βRI inhibitor (SB431542) was used to pretreat the cells. PBM irradiation resulted in a significant down-regulation of cell surface markers CD44, CD90, CD106, CD117, and Stro-1, indicating that the stem cells had undergone a differentiation program and had lost some of their state of stemness. Interestingly, PBM irradiation or TGF-1 directly were able to significantly down-regulate CD106, but not CD44 or 117, in a variety of mesenchymal stem cell lines. This modulation could be prevented by pre-incubation with antioxidants (NAC) or TGF-β inhibitors (SB431542 and SIS3) prior to PBM irradiation. Along with the loss of stem cell markers, PBM-treated cells exhibited a concurrent increase in expression of dentin differentiation markers, namely alkaline phosphatase and the dentin matrix markers dentin matrix protein 1 (DMP1), dentin sialoprotein (DSP), and osteopontin (OPN). These results indicate that PBM activation of TGF-β1 can direct hDSCs differentiation to dentin matrix producing cells that could contribute to the dentin regeneration.

7.3 PBM and tooth regeneration

Arany *et al* [32] explored the ability of PBM to induce dentin repair in a rodent tooth pulp capping model. To confirm whether PBM could promote dentin regeneration, PBM irradiation of the exposed pulp in the rat maxillary first molar was performed and reparative dentin induction was assessed with high-resolution microcomputed tomography. Increased dentin volumes at 12 weeks post-PBM irradiation as compared to non-irradiated controls were observed. Histological evaluation with biochemical staining demonstrated increased reparative dentin induced by PBM. As positive controls in these experiments, a calcium hydroxide dressing was used to induce production of dentin. Tertiary or reparative dentin is characterized by its distinct mineral composition and anatomical location. The PBM-induced reparative dentin was assessed with energy dispersive spectroscopy revealing an intermediate mineral content between adjacent normal dentin and pulp matrix. Further, Raman microscopy was performed to characterize the composition of normal tooth mineralized tissues including dentin, cementum, enamel, and alveolar bone. Raman analyses demonstrated the lower matrix (CH, 1450 cm^{-1}) and higher phosphate (PO_4^{3-}, 970 cm^{-1}) content of PBM-induced reparative dentin compared to adjacent native dentin that was deposited within and along the pulp walls in PBM-treated teeth.

7.4 TGF-β mediates PBM induction of dentin repair *in vivo*

Considering that *in vitro* studies had established a role for PBM-activated TGF-β1 in dentin differentiation, the role of TGF-β1 in PBM-induced dentin regeneration *in vivo* was examined. Two distinct interventional strategies were chosen to address this question. First, a chemical biology approach was pursued that utilized a controlled delivery of TGF-β reagents via PLG microspheres in the rat pulp-capping model to assess dental stem cell differentiation and mineralized dentin repair induction. As the *in vitro* studies had indicated that PBM irradiation could direct dental stem cell differentiation, we first established that the rat pulp possessed a comparable dental stem cell population, as human DSCs, and would respond similarly to PBM treatment. Rat tooth has DSCs as demonstrated by their cell surface marker positive for CD44, CD106, and CD117, and negative for CD45. PBM irradiation of rat pulp decreased expression of these markers at seven days, consistent with PBM effects on hDSCs *in vitro*. Controlled release of TGF-β1 directly via PLG microspheres also resulted in a similar decreased expression of these markers. Further, sustained release of a small-molecule inhibitor against TGF-βRI, SB431542, was achieved with PLG microspheres. Microspheres were first placed on the exposed pulp overnight to block TGF-β responsiveness in the pulp cell population. In one group, the pulp was re-exposed, and PBM irradiation was performed followed by reinsertion of these inhibitor microspheres. The presence of the TGF-β inhibitor prevented the down-regulation of dental stem cell surface markers observed with PBM irradiation. The use of a TGF-β neutralizing polyclonal antibody, 1D11, was able to similarly retain the of expression DSC markers following PBM irradiation. As PBM leads to a robust increase in reparative dentin, the role of TGF-β was also probed using controlled release of the TGF-βRI inhibitor, SB431542. Rat pulp was treated with either SB431542 alone, or with PBM irradiation combined with SB431542, and assessed with histology and high-resolution microcomputed tomography at 12 weeks. Rat teeth demonstrated minimal reparative dentin induction in both groups. A slight decrease in dentin in the PBM and SB431542 inhibitor group compared to the SB431542 inhibitor group alone was noted, but this was not statistically significant.

A second approach employing transgenic mice was used to confirm the role of TGF-β1 in PBM dentin induction. TGF-β responsiveness of pulp–dentin cells was designed to be interrupted by generation of conditional knockout (coKO) mice by crossing a DSPP^{Cre} with TGF-βRII$^{f/f}$. The gene dentin sialophosphoprotein (DSPP) is the most abundant non-collagenous protein present in dentin, encoding two distinct matrix proteins, dentin sialoprotein (DSP) and dentin phosphoprotein (DPP) [34, 35]. The key role of DSPP in dentin mineralization is highlighted by the occurrence of severe defects in mice and human dentin as a result of mutations or deletions in the DSPP gene [36, 37]. The TGF-βRII receptor is specific for TGF-β ligands, and has very high affinity for TGF-β isoforms 1, 2, and 3. TGF-βRII knockout mice have a lethal systemic inflammatory phenotype closely resembling the TGF-β knockouts and therefore cannot be produced, supporting the key role of TGF-βRII in normal pathophysiology [38, 39]. The cells of the pulp–dentin complex expressing DSPP include the dental stem cells, pre-odontoblasts, and mature odontoblasts that are all capable of potentially responding to PBM-mediated reparative induction of dentin [40]. Hence, the coKO mice will allow the targeting of most of the TGF-β responsive cells capable of inducing dentin within the pulp. Pulp exposure followed by either PBM irradiation and filling, or filling alone (control) in these coKO mice demonstrated minimal reparative dentin in both groups. Similar experiments in TGF-βRII$^{f/f}$ were observed to have the usual PBM-induced reparative dentin response. Taken together, data from both these *in vivo* approaches designed to inhibit TGF-β signaling indicate that TGF-β plays a key role in mediating PBM induction of dentin regeneration.

7.5 Conclusion

Regenerative medicine currently depends predominantly on *ex vivo* manipulated cells or delivery of exogenous factors such as recombinant growth factors [41–43]. These approaches, however, have significant limitations in terms of availability, costs, complexity, and potential side effects and toxicity. The results discussed in this chapter suggest a minimally invasive treatment, PBM irradiation, can be used to activate endogenous cues that can drive tooth regeneration bypassing the need to deliver exogenous cells or factors. Given the broad range of roles ROS and TGF-β play *in vivo*, the PBM effects noted in this study could be potentially harnessed to promote regeneration of various tissues such as bone, nerves, and muscle.

References

[1] Discher D E, mooney D J and Zandstra P W 2009 Growth factors, matrices, and forces combine and control stem cells. *Science* **324** 1673–7
[2] Xu Y, Shi Y and Ding S 2008 A chemical approach to stem-cell biology and regenerative medicine *Nature* **453** 338–44
[3] Suzuki T *et al* 2011 Induced migration of dental pulp stem cells for *in vivo* pulp regeneration *J. Dent. Res.* **90** 1013–8

[4] Halliwell B and Gutteridge J M 1984 Oxygen toxicity, oxygen radicals, transition metals and disease *Biochem. J.* **219** 1–14

[5] Karu T I *et al* 2005 Absorption measurements of a cell monolayer relevant to phototherapy: reduction of cytochrome c oxidase under near IR radiation *J. Photochem. Photobiol.* B **81** 98–106

[6] Chen A C-H *et al* 2009 Role of reactive oxygen species in low level light therapy *Proc. SPIE* 7165 716502-1

[7] Huang Y Y *et al* 2012 Low-level laser therapy (LLLT) reduces oxidative stress in primary cortical neurons *in vitro J. Biophotonics*

[8] Hirata S *et al* 2010 Low-level laser irradiation enhances BMP-induced osteoblast differentiation by stimulating the BMP/Smad signaling pathway *J. Cell. Biochem.* **111** 1445–52

[9] Arany P R *et al* 2007 Activation of latent TGF-beta1 by low-power laser *in vitro* correlates with increased TGF-beta1 levels in laser-enhanced oral wound healing *Wound Repair Regen* **15** 866–74

[10] Statius van Eps R G and LaMuraglia G M 1997 Photodynamic therapy inhibits transforming growth factor beta activity associated with vascular smooth muscle cell injury *J. Vasc. Surg.* **25** 1044–52 1052–3

[11] Byrnes, K R *et al* 2005 Light promotes regeneration and functional recovery and alters the immune response after spinal cord injury *Lasers Surg. Med.* **36** 171–85

[12] Toyokawa, H *et al* 2003 Promotive effects of far-infrared ray on full-thickness skin wound healing in rats. *Exp. Biol. Med.* **228** 724–9

[13] Roberts A B *et al* 1981 New class of transforming growth factors potentiated by epidermal growth factor: isolation from non-neoplastic tissues *Proc. Natl Acad. Sci. USA* **78** 5339–43

[14] Cheifetz S *et al* 1987 The transforming growth factor-beta system, a complex pattern of cross-reactive ligands and receptors *Cell* **48** 409–15

[15] Moses H L *et al* 1981 Transforming growth factor production by chemically transformed cells *Cancer Res.* **41** 2842-8

[16] Watabe T and Miyazono K 2009 Roles of TGF-beta family signaling in stem cell renewal and differentiation *Cell Res.* **19** 103–15

[17] Pera M F and Tam P P 2010 Extrinsic regulation of pluripotent stem cells *Nature* **465** 713–20

[18] Nishimura E K *et al* 2010 Key roles for transforming growth factor beta in melanocyte stem cell maintenance *Cell Stem Cell* **6** 130–40

[19] Arany P R; mooney D J 2011 At the edge of translation—materials to program cells for directed differentiation *Oral Dis.* **17** 241–51

[20] Yamanaka S 2009 Elite and stochastic models for induced pluripotent stem cell generation *Nature* **460** 49–52

[21] Mullen A C *et al* 2011 Master transcription factors determine cell-type-specific responses to TGF-beta signaling *Cell* **147** 565–76

[22] Xi Q *et al* 2011 A poised chromatin platform for TGF-beta access to master regulators *Cell* **147** 1511–24

[23] Oka S *et al* 2007 Cell autonomous requirement for TGF-beta signaling during odontoblast differentiation and dentin matrix formation *Mech. Dev.* **124** 409–15

[24] D'Souza R N *et al* 1998 TGF-beta1 is essential for the homeostasis of the dentin–pulp complex *Eur. J. Oral Sci.* **106** 185–91

[25] Gronthos S *et al* 2000 Postnatal human dental pulp stem cells (DPSCs) *in vitro* and *in vivo Proc. Natl. Acad. Sci. USA* **97** 13625–30

[26] Iohara K *et al* 2006 Side population cells isolated from porcine dental pulp tissue with self-renewal and multipotency for dentinogenesis, chondrogenesis, adipogenesis, and neurogenesis *Stem Cells* **24** 2493–503

[27] Braut A, Kollar E J and Mina M 2003 Analysis of the odontogenic and osteogenic potentials of dental pulp *in vivo* using a Col1a1-2.3-GFP transgene *Int. J. Dev. Biol.* **47** 281–92

[28] Sakai V T *et al* 2010 SHED differentiate into functional odontoblasts and endothelium *J. Dent. Res.* **89** 79–6

[29] Seo B M *et al* 2004 Investigation of multipotent postnatal stem cells from human periodontal ligament *Lancet* **364** 149–55

[30] Sonoyama W *et al* 2008 Characterization of the apical papilla and its residing stem cells from human immature permanent teeth: a pilot study *J. Endod.* **34** 166–71

[31] Tziafas D, Smith A J and Lesot H 2000 Designing new treatment strategies in vital pulp therapy *J. Dent.* **28** 77–92

[32] Arany P R *et al* 2014 Photoactivation of endogenous latent transforming growth factor-beta1 directs dental stem cell differentiation for regeneration *Sci. Transl. Med.* **6** 238–69

[33] Yu J *et al* 2010 Differentiation potential of STRO-1+ dental pulp stem cells changes during cell passaging *BMC Cell Biol.* **11** 32

[34] MacDougall M *et al* 1997 Assignment of dentin sialophosphoprotein (DSPP) to the critical DGI2 locus on human chromosome 4 band q21.3 by *in situ hybridization Cytogenet Cell Genet* **79** 121–2

[35] Ritchie H H and Wang L H 1996 Sequence determination of an extremely acidic rat dentin phosphoprotein *J. Biol. Chem.* **271** 21695–8

[36] Sreenath T *et al* 2003 Dentin sialophosphoprotein knockout mouse teeth display widened predentin zone and develop defective dentin mineralization similar to human dentinogenesis imperfecta type III *J. Biol. Chem.* **278** 24874–80

[37] McKnight D A *et al* 2008 A comprehensive analysis of normal variation and disease-causing mutations in the human DSPP gene *Hum. Mutat.* **29** 1392–404

[38] Leveen P *et al* 2002 Induced disruption of the transforming growth factor beta type II receptor gene in mice causes a lethal inflammatory disorder that is transplantable *Blood* **100** 560–8

[39] Wrana J L *et al* 1992 TGF beta signals through a heteromeric protein kinase receptor complex *Cell* **71** 1003–14

[40] Sreenath T L *et al* 1999 Spatial and temporal activity of the dentin sialophosphoprotein gene promoter: differential regulation in odontoblasts and ameloblasts *Int. J. Dev. Biol.* **43** 509-16

[41] Mustoe T A *et al* 1989 Reversal of impaired wound healing in irradiated rats by platelet-derived growth factor-BB *Am. J. Surg.* **158** 345–50

[42] McKay W F, Peckham S M and Badura J M 2007 A comprehensive clinical review of recombinant human bone morphogenetic protein-2 (INFUSE Bone Graft) *Int. Orthop.* **31** 729–34

[43] Naldini L 2011 *Ex vivo* gene transfer and correction for cell-based therapies *Nat. Rev. Genet.* **12** 301–15

IOP Concise Physics

Photomedicine and Stem Cells

The Janus face of photodynamic therapy (PDT) to kill cancer stem cells, and photobiomodulation
(PBM) to stimulate normal stem cells

Heidi Abrahamse and Michael R Hamblin

Chapter 8

The role of stem cells and progenitor cells in PBM for brain disorders

8.1 Introduction

Although in the past it was generally accepted that the adult central nervous system (CNS) could not repair itself, recent discoveries in the area of neuronal stem cells and neuroprogenitor cells have brought this outdated dogma into question [1]. The discovery of BrdU (an exogenously administered marker that can be injected into mice or added to cell cultures) that efficiently labels proliferating cells and can easily be detected by antibodies in tissue sections has led to proliferating neurons being identified in the brains of adult laboratory animals [2] as well as humans [3]. In mice, there are thought to be two sources of neural stem cells (NSCs) and their offspring, neural progenitor cells (NPCs). In developing brains neurogenesis occurs in so-called 'neurogenic niches' located in the subventricular zone (SVZ) of the lateral ventricles and in the subgranular layer (SGL) of the hippocampal dentate gyrus (DG) [4]. These regions can be thought of as specialized niches for NSC [5]. In rodents NSC produce neuroblasts that migrate from the SVZ along a discrete pathway, the rostral migratory stream, into the olfactory bulb where they form mature neurons involved in the sense of smell. New SGL neurons migrate only a short distance and differentiate into hippocampal granule cells.

In the developing brain, neurogenesis is more widespread occurring in the ventricular and subventricular proliferative zones. Neuroprogenitor cells can divide to either produce neurons or glia, depending on the intrinsic and environmental cues. Neurogenic niches are characterized by a comparatively high vascular density and, in many cases, interaction with the cerebrospinal fluid (CSF) [6]. Both the vasculature and the CSF represent a source of signaling molecules, which can be relatively rapidly modulated by external factors and circulated through the CNS. As the brain develops, there is vascular remodeling and a compartmentalization and dynamic modification of the ventricular surface, which may be responsible for the change in the proliferative properties.

8.2 PBM and the brain

The changes in expression levels of proteins involved in antioxidant and redox-regulation, anti-apoptotic and pro-survival, cellular proliferation, etc, mean that distinct changes in tissue homeostasis, healing, and regeneration can be expected after photobiomodulation (PBM). For instance, structural proteins such as collagen are newly synthesized in order to repair tissue damage [7]. Cells at risk of dying in tissue that has been subjected to ischemic or other insults are protected [8]. Stem cells are activated to leave their niche, proliferate, and differentiate [9, 10]. Pain and inflammation are reduced [11]. Blood flow is increased [12] (possibly as a result of the release of NO [13]), which also stimulates lymphatic drainage, thereby reducing edema [14]. In addition to the foregoing, there are some PBM tissue mechanisms that are specific to the brain. One of the most important is an increase in cerebral blood flow often reported after transcranial photobiomodulation (tPBM) [15], leading to increased tissue oxygenation, and more oxidized CCO as measured by near-infrared (NIR) spectroscopy [16]. tPBM has been shown to reduce activated microglia in the brains of TBI mice as measured by IBA1 (ionized calcium-binding adapter molecule-1) expression thus demonstrating reduced neuroinflammation [17]. tPBM has been shown to increase neurogenesis (formation of new brain cells derived from neuroprogenitor cells) [18], and synaptogenesis (formation of new connections between existing brain cells) [19] both in TBI mice. Figure 8.1 shows a graphical representation of a variety of these brain-specific tissue mechanisms.

8.3 PBM for stroke

One of the most well-investigated application of PBM to the brain, lies in its possible use as a treatment for acute stroke [20]. Animal models such as rats and rabbits were first used as laboratory models, and these animals had experimental strokes induced by a variety of methods, and were then treated with light (usually 810 nm laser) within 24 h of stroke onset [21]. In these studies, intervention by tPBM within 24 h had meaningful beneficial effects. For the rat models, stroke was induced by middle cerebral artery occlusion (MCAO) via an insertion of a filament into the carotid artery or via craniotomy [22, 23]. Stroke induction in the 'rabbit small clot embolic model' (RSCEM) was by injection of a preparation of small blood clots (made from blood taken from a second donor rabbit) into a catheter placed in the right internal carotid artery [24]. These studies and the treatments and results are listed in table 8.1.

The highly encouraging effects of PBM in animal models of stroke encouraged a series of three clinical trials to be conducted. These trials were called 'Neurothera Effectiveness and Safety Trials' (NEST-1 [28], NEST-2 [29], and NEST-3 [30]) and used an 810 nm laser applied to the shaved head within 24 h of patients suffering an ischemic stroke. The first study, NEST-1, enrolled 120 patients between the ages of 40 to 85 years of age with a diagnosis of ischemic stroke involving a neurological

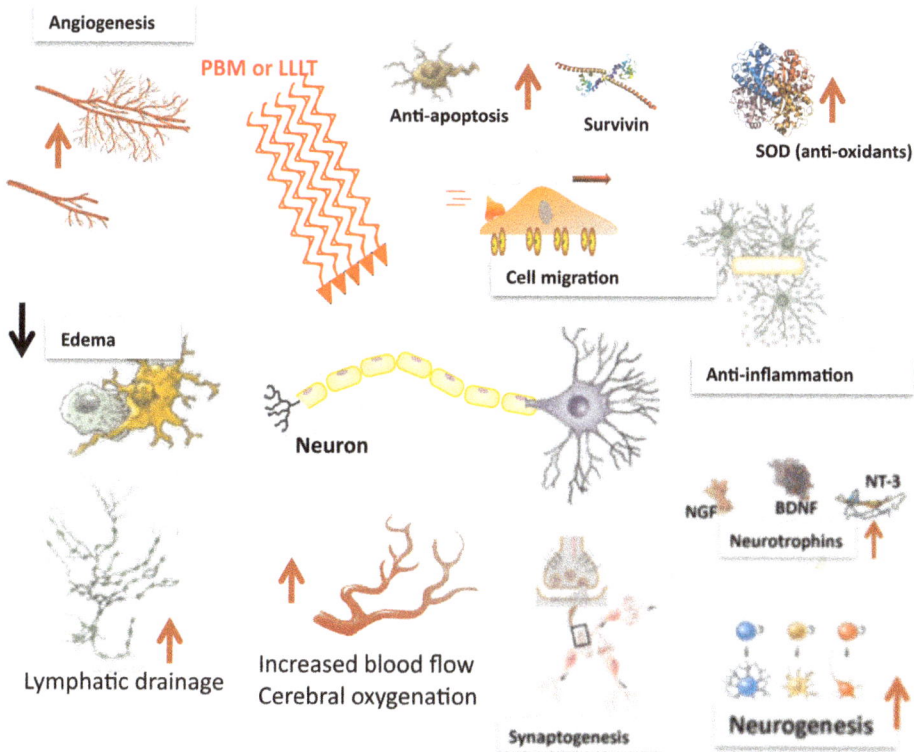

Figure 8.1. Brain-specific mechanisms of tPBM. PBM can lead to decreases in neuronal apoptosis and excitotoxicity and lessening of inflammation and reduction of edema due to increased lymphatic flow which, together with protective factors such as antioxidants, will all help to reduce progressive brain damage. Increases in angiogenesis, expression of neurotrophins leading to activation of NPCs and more cell migration, and increased synaptogenesis may all contribute to the brain repairing itself from damage sustained in the trauma.

deficit that could be measured. The purpose of this first clinical trial was to demonstrate the safety and effectiveness of laser therapy for stroke within 24 h [28]. tPBM significantly improved outcome in human stroke patients, when applied at ~18 h post-stroke, over the entire surface of the head (20 points in the 10/20 EEG system) regardless of stroke [28]. Only one laser treatment was administered, and 5 days later, there was significantly greater improvement in the real but not in the sham-treated group (p <.05, NIH Stroke Severity Scale). This significantly greater improvement was still present at 90 days post-stroke, where 70% of the patients treated with real-PBM had a successful outcome, while only 51% of sham controls did. The second clinical trial, NEST-2, enrolled 660 patients, aged 40 to 90, who were randomly assigned to one of two groups (331 to PBM, 327 to sham) [31]. Beneficial results (p < .04) were found for the moderate and moderate-severe (but not for the severe) stroke patients, who received the real laser protocol [31]. These results suggested that the overall severity of the individual stroke should be taken into consideration in future studies, and very severe patients are unlikely to recover

Table 8.1. Reports of tPBM used for stroke in animal models. CW, continuous wave; LLLT, low-level light therapy; MCAO, middle cerebral artery occlusion; NOS, nitric oxide synthase; RSCEM, rabbit small clot embolic model; TGF-β1, transforming growth factor β1.

Subject	Stroke model	Parameters	Effect	References
Rat	MCAO	660 nm, 8.8 mW, 2.64 J cm^{-2}, pulse frequency of 10 kHz, laser applied at cerebrum at 1, 5, and 10 min	Suppression of NOS activity and up regulation of TGF-β1	[22]
Rat	MCAO	808 nm, 7.5 mW cm^{-2}, 0.9 J cm^{-2}, 3.6 J cm^{-2} at cortical surface, CW and pulse wave at 70 Hz, 4 mm diameter	Administration of PBM 24 h after stroke onset induced functional benefit and neurogenesis induction	[23]
Rabbit	RSCEM	808 nm +/− 5 nm, 7.5 W cm^{-2}, 2 min duration 3 h after stroke and 25 mW cm^{-2} 10 min duration 1 or 6 h after stroke	Improved behavioral performance and durable effect after PBM within 6 h from stroke onset	[25]
Rat	MCAO	808 nm, 0.5 mW cm^{-2}, 0.9 J cm^{-2} on brain 3 mm dorsal to the eye and 2 mm anterior to the ear	PBM applied at different locations on the skull improved neurological function after acute stroke	[26]
Rabbit	RSCEM	808 nm, 7.5 mW cm^{-2}, 0.9 J cm^{-2}, 3.6 J cm^{-2} at cortical surface, CW, 300 min, pulse at 1 kHz, 2 min at 100 Hz	PBM administered 6 h after embolic stroke resulted in clinical improvements in rabbits	[27]

with any kind of treatment. The last clinical trial, NEST-3, was planned for 1000 patients. Patients in this study were not to receive tissue plasminogen activator, but the study was prematurely terminated by the DSMB for futility (an expected lack of statistical significance) [30]. NEST1 was considered successful, even though as a phase 1 trial, it was not designed to show efficacy. NEST2 was partially successful when the patients were stratified to exclude very severe strokes or strokes deep within the brain [29]. However, the failure of NEST3 may have been due to the decision to use only one tPBM treatment, instead of a series of treatments. Moreover, the optimum brain areas to be treated in acute stroke remain to be determined.

8.4 PBM for traumatic brain injury

There have been a number of studies looking at the effects of PBM in animal models of TBI. Oron's group was the first [32] to demonstrate that a single exposure of the mouse head to a NIR laser (808 nm) a few hours after creation of a TBI lesion could improve neurological performance and reduce the size of the brain lesion. A weight-drop device was used to induce a closed-head injury in the mice. An 808 nm diode laser with two energy densities ($1.2–2.4$ J cm^{-2} over 2 min of irradiation with 10 and 20 mW cm^{-2}) was delivered to the head 4 h after TBI was induced. Neurobehavioral function was assessed by the neurological severity score (NSS). There were no significant differences in NSS between the power densities (10 versus 20 mW cm^{-2}) or significant differentiation between the control and laser-treated group at early time points (24 and 48 h) post TBI. However, there was a significant improvement (27% lower NSS score) in the PBM group at times of 5 days to 4 weeks. The laser-treated group also showed a smaller loss of cortical tissue than the sham group [32]

Hamblin's laboratory then went on (in a series of papers [32]) to show that a 810 nm laser (and 660 nm laser) could benefit experimental TBI both in a closed-head weight drop model [33], and also in controlled cortical impact model in mice [34]. Wu *et al* [33] explored the effect that varying the laser wavelengths of PBM had on closed-head TBI in mice. Mice were randomly assigned to PBM-treated group or to sham group as a control. Closed-head injury (CHI) was induced via a weight drop apparatus. To analyze the severity of the TBI, the NSS was measured and recorded. The injured mice were then treated with varying wavelengths of laser (665, 730, 810, or 980 nm) at an energy level of 36 J cm^{-2} at 4 h directed onto the scalp. The 665 nm and 810 nm groups showed significant improvement in NSS when compared to the control group at day 5 to day 28. Results are shown in figure 8.2. Conversely, the 730 and 980 nm groups did not show a significant improvement in NSS and these wavelengths did not produce similar beneficial effects as in the 665 nm and 810 nm PBM groups [33]. The tissue chromophore cytochrome c oxidase (CCO) is proposed to be responsible for the underlying mechanism that produces the many PBM effects that are the byproduct of PBM. CCO has absorption bands around 665 nm and 810 nm, while it has low absorption bands at the wavelength of 730 nm [35]. It should be noted that this particular study found that the 980 nm wavelength did not produce the same positive effects as the 665 nm and 810 nm wavelengths did; nevertheless previous studies did find that the 980 nm wavelength was an active one for PBM. Wu *et al* proposed these dissimilar results may be due to the variance in the energy level, irradiance, etc, between the other studies and this particular study [33].

Ando *et al* [34] used the 810 nm wavelength laser parameters from the previous study and varied the pulse modes of the laser in a mouse model of TBI. These modes consisted of either pulsed wave at 10 Hz or at 100 Hz (50% duty cycle) or continuous wave laser. For the mice, TBI was induced with a controlled cortical impact device via open craniotomy. A single treatment with an 810 nm Ga–Al–As diode laser with a power density of 50 mW m^{-2} and an energy density of 36 J cm^{-2} was given via tPBM to the closed head in mice for a duration of 12 min at 4 h post controlled cortical impact (CCI). At 48 h to 28 days post TBI, all laser-treated groups had significant decreases in the measured NSS when compared to the control (figure 8.3). Although all laser-treated groups had similar NSS improvement rates up to day 7, the pulse wave (PW) (laser) 10 Hz group began to show greater improvement beyond this point. At day 28, the forced swim test for depression and anxiety was

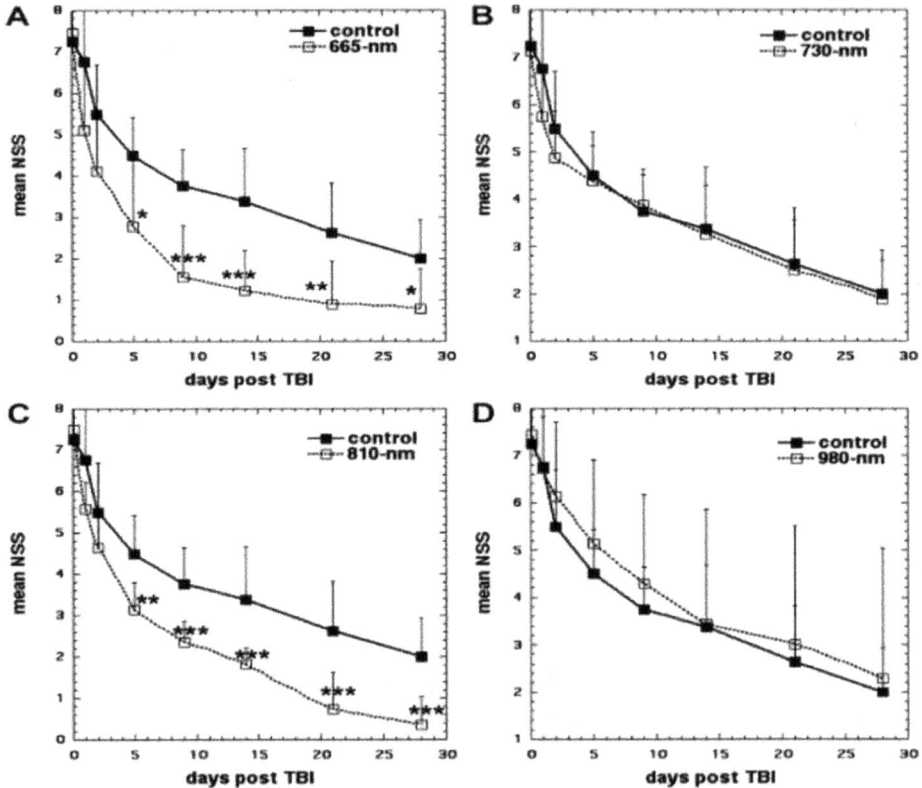

Figure 8.2. Effect of different laser wavelengths of tPBM in closed-head TBI in mice. (A) Sham-treated control versus 665 nm laser. (B) Sham-treated control versus 730 nm laser. (C) Sham-treated control versus 810 nm laser. (D) Sham-treated control versus 980 nm laser. Points are means of 8–12 mice and bars are SD. $*p < 0.05$; $**p < 0.01$; $***p < 0.001$ (one-way ANOVA). Reprinted with permission from [36].

used and showed a significant decrease in the immobility time for the PW 10 Hz group. In the tail suspension test, which measures depression and anxiety, there was also a significant decrease in the immobility time at day 28, and this time also at day 1, in the PW 10 Hz group.

Studies using immunofluorescence of mouse brains showed that tPBM increased neuroprogenitor cells in the DG (figure 8.4) and SVZ at 7 days after treatment [38]. These neuroprogenitor cells were detected by injection of BrdU (into the mice) each day for 5 days before sacrifice which led to it being incorporated into dividing cells in the brain [18]. The neurotrophin called brain derived neurotrophic factor (BDNF) was also increased in the DG and SVZ at 7 days, while the marker (synapsin-1) for synaptogenesis and neuroplasticity was increased in the cortex at 28 days, but not in the DG, SVZ at 28 days, or at any location at 7 days [19] Learning and memory as measured by the Morris water maze was also improved by tPBM [18].

Whalen's laboratory [17] and Whelan's laboratory [39] also successfully demonstrated therapeutic benefits of tPBM for TBI in mice and rats, respectively. Zhang *et al* [40] showed that secondary brain injury occurred to a worse degree in mice that

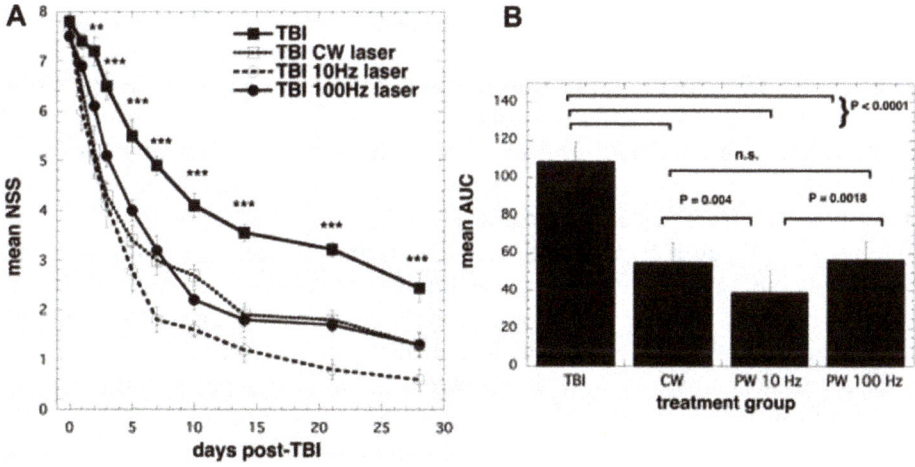

Figure 8.3. Effects of pulsing in tPBM for CCI-TBI in mice. (A) Time course of NSS of mice with TBI receiving either control (no laser treatment), or 810 nm laser treatment (36 J cm^{-2} delivered at 50 mW cm^{-2} with a spot size of 0.78 cm^2 in either CW, PW 10 Hz, or PW 100 Hz modes). Results are expressed as mean +/− SEM ***$p < 0.001$ versus the other conditions. (B) Mean areas under the NSS–time curves in the two-dimensional coordinate system over the 28 day study for the four groups of mice. Results are means +/− SD ($n = 10$). Reprinted from [37] (open access).

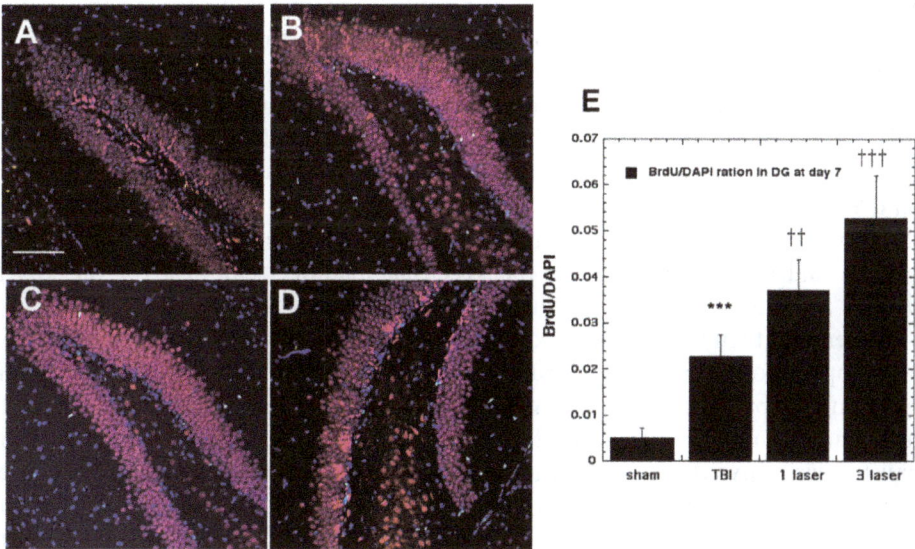

Figure 8.4. BrdU–NeuN double-stained images and analysis in the neurogenic region of the hippocampus (DG) at 7 days post-TBI. (A) Sham, (B) CCI-TBI, (C) 1xtLLLT, (D) 3xtLLLT, (E) mean BrdU/DAPI ratios ± SD ($n = 5$); normalization of the readings was done using BrdU versus DAPI (labeling the nuclei). Scale bar 100 μm. ***$p < 0.001$ versus sham; †† $p < 0.01$ versus TBI; ††† p < 0.001 versus TBI.

had been genetically engineered to lack the 'immediate early response' gene X-1 (IEX-1) when exposed to a gentle head impact (this injury is thought to closely resemble mild TBI (mTBI) in humans). Exposing IEX-1 knockout mice to PBM 4 h post-injury suppressed proinflammatory cytokine expression of interleukin (IL)-Iβ and IL-6, but up-regulated TNF-α. The lack of IEX-1 decreased adenosine triphosphate (ATP) production, but exposing the injured brain to PBM elevated ATP production back to near normal levels.

Dong *et al* [41] even further improved the beneficial effects of PBM on TBI in mice, by combining the treatment with metabolic substrates such as pyruvate and/or lactate. The goal was to even further improve mitochondrial function. This combinatorial treatment was able to reverse memory and learning deficits in TBI mice back to normal levels, as well as leaving the hippocampal region completely protected from tissue loss; a stark contrast to that found in control TBI mice that exhibited severe tissue loss from secondary brain injury.

Margaret Naeser and collaborators have tested PBM in human subjects who had suffered TBI in the past [42]. Many sufferers of severe or even moderate TBI have very long lasting and even life-changing sequelae (headaches, cognitive impairment, and difficulty sleeping) that prevent them working or living any kind or normal life. These individuals may have been high achievers before the accident that caused damage to their brain [43]. Initially Naeser published a report [44] describing two cases she treated with PBM applied to the forehead twice a week. A 500 mW continuous wave light-emitting diode (LED) source (a mixture of 660 nm red and 830 nm NIR LEDs) with a power density of 22.2 mW cm^{-2} (area of 22.48 cm^2), was applied to the forehead for a typical duration of 10 min (13.3 J cm^{-2}). In the first case study the patient reported that she could concentrate on tasks for a longer period of time (the time able to work at a computer increased from 30 min to 3 h). She had a better ability to remember what she read, decreased sensitivity when receiving haircuts in the spots where PBM was applied, and improved mathematical skills after undergoing PBM. The second patient had statistically significant improvements compared to prior neuropsychological tests after 9 months of treatment. The patient had a 2 standard deviation (SD) increase on tests of inhibition and inhibition accuracy (9th percentile to 63rd percentile on the Stroop test for executive function and a 1 SD increase on the Wechsler memory scale test for the logical memory test, 83rd percentile to 99th percentile) [45].

Naeser *et al* then went on to report a case series of a further 11 patients [46]. This was an open protocol study that examined whether scalp application of red and NIR light could improve cognition in patients with chronic mTBI. This study had 11 participants ranging in age from 26 to 62 (6 males, 5 females) who suffered from persistent cognitive dysfunction after mTBI. The participants' injuries were caused by motor vehicle accidents, sports-related events, and for one participant an improvised explosive device (IED) blast. The tPBM consisted of 18 sessions (Monday, Wednesday, and Friday for 6 weeks) and commenced anywhere from 10 months to 8 years post-TBI. A total of 11 LED clusters (5.25 cm in diameter, 500 mW, 22.2 mW cm^{-2}, 13 J cm^{-2}) were applied for about 10 min per session (5 or 6 LED placements per set, Set A and then Set B, in each session). Neuropsychological testing was performed pre-LED application and 1 week, 1 month, and 2 months after the final treatment. Naeser and colleagues found that there was a significant positive linear trend observed for the Stroop test for executive function, in trial 2 inhibition ($p = 0.004$); Stroop, trial 4 inhibition switching ($p = 0.003$); California Verbal Learning Test (CVLT)-II, total trials 1–5 ($p = 0.003$); and CVLT-II, long delay free recall ($p = 0.006$). Improved sleep and fewer post-traumatic stress disorder (PTSD) symptoms, if present beforehand, were observed after treatment. Participants and family members also reported better social function and a better ability to perform interpersonal and occupational activities. Although these results were significant, further placebo-controlled studies will be needed to ensure the reliability of this these data [46]. Henderson and Morries [47] used a high-power NIR laser (10–15 W at 810 and 980 nm) applied to the head to treat a patient with moderate TBI. The patient received 20 NIR applications over a 2 month period. They carried out anatomical magnetic resonance imaging (MRI) and perfusion single-photon emission computed tomography (SPECT). The patient showed decreased depression, anxiety, headache, and insomnia, and cognition and quality of life improved, accompanied by changes in the SPECT imaging.

8.5 PBM/LLLT for other brain disorders

Although the role of PBM in stimulating stem cells in the brain has been mostly studied in stroke and TBI, the treatment of many other brain disorders is beginning to be investigated using various kinds of PBM approaches. These new applications can be divided into two broad groups, neurodegenerative diseases, and psychiatric disorders.

The neurodegenerative diseases that have been treated with PBM include Alzheimer's disease (AD), Parkinson's disease (PD), and primary progressive aphasia.

A convincing study [48] on AD was carried out in an AβPP transgenic mouse. tPBM (810 nm laser) was administered at different doses 3 times/week for 6 months starting at 3 months of age. The numbers of Aβ plaques were significantly reduced in the brain with administration of tPBM in a dose-dependent fashion. tPBM mitigated the behavioral effects seen with advanced amyloid deposition and reduced the expression of inflammatory markers in the transgenic mice. Clinical studies on AD have been recently carried out in Canada with great success, but in a very small trial (five patients) [49].

PD has been extensively investigated in animal models by the Mitrofanis laboratory in Australia [50, 51]. These workers have demonstrated the effectiveness of PBM in several different animal models of PD, including mice who had been administered 1-methyl-4-phenyl-1,2,3,6-tetrahydropyridine (MPTP). MPTP is converted into to the neurotoxin MPP+, which causes permanent symptoms of PD by destroying dopaminergic neurons in the substantia nigra of the brain [52]. Mice were treated with tPBM (670 nm LED, 40 mW cm^{-2}, 3.6 J cm^{-2}) 15 min after each MPTP injection repeated 4 times over 30 h. There were significantly more (35%–45%) dopaminergic cells in the brains of the tPBM-treated mice [53]. A subsequent study showed similar results in a chronic mouse model of MPTP-induced PD [54]. They repeated their studies in another mouse model of PD, the tau transgenic mouse strain (K3) that has a progressive degeneration of dopaminergic cells in the substantia nigra pars compacta (SNc) [55]. They went on to test a surgically implanted intracranial fiber designed to deliver either 670 nm LED (0.16 mW) or 670 nm laser (67 mW) into the lateral ventricle of the brain in MPTP-treated mice [56]. Both low-power LEDs and a high-power laser were effective in preserving SNc cells, but the laser was considered to be unsuitable for long term use (6 days) due to excessive heat production. These authors also reported a protective effect of abscopal light exposure (head shielded) in this mouse model [57]. Yet another model they used was rats treated with 6-hydroxydopamaine [58]. Recently this group has tested their implanted fiber approach in a model of PD in adult Macaque monkeys treated with MPTP [59]. Clinical evaluation of Parkinson's symptoms (posture, general activity, bradykinesia, and facial expression) in the monkeys were improved at low doses of light (24 J or 35 J) compared to high doses (125 J) [60].

The psychiatric disorders that have been treated so far have chiefly been major depression and anxiety. A common and well-accepted animal model of depression is called 'chronic mild stress' [61]. After exposure to a series of chronic unpredictable mild stressors, animals develop symptoms seen in human depression, such as anhedonia (loss of the capacity to experience pleasure, a core symptom of major depressive disorder), weight loss or slower weight gain, decrease in locomotor activity, and sleep disorders [62]. Wu et al used Wistar rats to show that after 5 weeks of chronic stress, application of tPBM 3 times a week for 3 weeks (810 nm laser, 100 Hz with 20% duty cycle, 120 J cm^{-2}) gave significant improvement in the forced swimming test (FST) [63]. In a similar study Salehpour et al [64] compared the effects of two different lasers (630 nm at 89 mW cm^{-2}, and 810 nm at 562 mW cm^{-2}, both pulsed at 10 Hz, 50% duty cycle). The 810 nm laser proved better than the 630 nm laser in the FST and in the elevated plus maze, and also reduced blood cortisol levels. The first clinical study on depression and anxiety was published by Schiffer et al in 2009 [65]. They used a fairly small-area 1 W 810 nm LED array applied to the forehead in patients with major depression and anxiety. They found improvements in the Hamilton depression rating scale (HAM-D) and the Hamilton anxiety rating scale (HAM-A) 2 weeks after a single treatment. They also found increases in frontal pole regional cerebral blood flow (rCBF) during the light delivery using a commercial NIR spectroscopy device. Cassano and co-workers [66] used tPBM with an 810 nm laser (700 mW cm^{-2} and a fluence of 84 J cm^{-2} delivered per session) for 6 sessions in patients with major depression. Baseline mean HAM-D17 scores decreased from 19.8 ± 4.4 (SD) to 13 ± 5.35 (SD) after treatment ($p = 0.004$).

8.6 Conclusion

The relevance of neural stem cells or neuroprogenitor cells to the dramatic improvements PBM can make to the brain remains uncertain at the present time. Nevertheless, the fact that BrdU positive cells in the brain have been demonstrated after PBM suggests that they will turn out to be important. The ever-increasing list of brain disorders that have been attributed (at least partly) to defects in neurogenesis in the brain continues to grow. Likewise, the list of brain disorders that have been improved by PBM also continues to grow.

References

[1] Yamashima T, Tonchev A B and Yukie M 2007 Adult hippocampal neurogenesis in rodents and primates: endogenous, enhanced, and engrafted *Rev. Neurosci.* **18** 67–82

[2] Landgren H and Curtis M A 2011 Locating and labeling neural stem cells in the brain *J. Cell. Physiol.* **226** 1–7

[3] Crespel A *et al* 2005 Increased number of neural progenitors in human temporal lobe epilepsy *Neurobiol. Dis.* **19** 436–50

[4] Lennington J B, Yang Z and Conover J C 2003 Neural stem cells and the regulation of adult neurogenesis *Reprod. Biol. Endocrinol.* **1** 99

[5] Lim D A, Huang Y C and Alvarez-Buylla A 2007 The adult neural stem cell niche: lessons for future neural cell replacement strategies *Neurosurg. Clin. N. Am.* **18** 81–92 ix

[6] Stolp H B and Molnar Z 2015 Neurogenic niches in the brain: help and hindrance of the barrier systems *Front. Neurosci.* **9** 20

[7] Tatmatsu-Rocha J C *et al* 2016 Low-level laser therapy (904 nm) can increase collagen and reduce oxidative and nitrosative stress in diabetic wounded mouse skin *J. Photochem. Photobiol.* B **164** 96–102

[8] Sussai D A *et al* 2010 Low-level laser therapy attenuates creatine kinase levels and apoptosis during forced swimming in rats *Lasers Med. Sci.* **25** 115–20

[9] Oron A and Oron U 2016 Low-level laser therapy to the bone marrow ameliorates neurodegenerative disease progression in a mouse model of alzheimer's disease: a minireview *Photomed. Laser. Surg.*

[10] Zhang Q *et al* 2016 Noninvasive low-level laser therapy for thrombocytopenia *Sci. Transl. Med.* **8** 349ra101

[11] Chow R T *et al* 2009 Efficacy of low-level laser therapy in the management of neck pain: a systematic review and meta-analysis of randomised placebo or active-treatment controlled trials *Lancet.* **374** 1897–908

[12] Samoilova K A *et al* 2008 Role of nitric oxide in the visible light-induced rapid increase of human skin microcirculation at the local and systemic levels: II. healthy volunteers *Photomed. Laser Surg.* **26** 443–9

[13] Mitchell U H and Mack G L 2013 Low-level laser treatment with near-infrared light increases venous nitric oxide levels acutely: a single-blind, randomized clinical trial of efficacy *Am. J. Phys. Med. Rehabil.* **92** 151–6

[14] Dirican A *et al* 2011 The short-term effects of low-level laser therapy in the management of breast-cancer-related lymphedema *Support Care Cancer.* **19** 685–90

[15] Salgado A S *et al* 2015 The effects of transcranial LED therapy (TCLT) on cerebral blood flow in the elderly women *Lasers Med. Sci.* **30** 339–46

[16] Rojas J C and Gonzalez-Lima F 2013 Neurological and psychological applications of transcranial lasers and LEDs *Biochem. Pharmacol.* **86** 447–57

[17] Khuman J *et al* 2012 Low-level laser light therapy improves cognitive deficits and inhibits microglial activation after controlled cortical impact in mice *J. Neurotrauma.* **29** 408–17

[18] Xuan W *et al* 2014 Transcranial low-level laser therapy enhances learning, memory, and neuroprogenitor cells after traumatic brain injury in mice *J. Biomed. Opt.* **19** 108003

[19] Xuan W *et al* 2015 Low-level laser therapy for traumatic brain injury in mice increases brain derived neurotrophic factor (BDNF) and synaptogenesis *J. Biophoton.* **8** 502–11

[20] Leung M C *et al* 2002 Treatment of experimentally induced transient cerebral ischemia with low energy laser inhibits nitric oxide synthase activity and up-regulates the expression of transforming growth factor-beta 1 *Lasers Surg. Med.* **31** 283–8

[21] Peplow P V 2015 Neuroimmunomodulatory effects of transcranial laser therapy combined with intravenous tPA administration for acute cerebral ischemic injury *Neural Regen. Res.* **10** 1186–90

[22] Oron A *et al* 2006 Low-level laser therapy applied transcranially to rats after induction of stroke significantly reduces long-term neurological deficits *Stroke.* **37** 2620–4

[23] Zhang L *et al* 2000 Quantitative measurement of motor and somatosensory impairments after mild (30 min) and severe (2 h) transient middle cerebral artery occlusion in rats *J. Neurol. Sci.* **174** 141–6

[24] Meyer D M, Chen Y and Zivin J A 2016 Dose-finding study of phototherapy on stroke outcome in a rabbit model of ischemic stroke *Neurosci. Lett.* **630** 254–8

[25] Lapchak P A *et al* 2007 Transcranial near-infrared light therapy improves motor function following embolic strokes in rabbits: an extended therapeutic window study using continuous and pulse frequency delivery modes *Neuroscience.* **148** 907–14

[26] Detaboada L *et al* 2006 Transcranial application of low-energy laser irradiation improves neurological deficits in rats following acute stroke *Lasers Surg. Med.* **38** 70–3

[27] Lapchak P A, Wei J and Zivin J A 2004 Transcranial infrared laser therapy improves clinical rating scores after embolic strokes in rabbits *Stroke.* **35** 1985–8

[28] Lampl Y *et al* 2007 Infrared laser therapy for ischemic stroke: a new treatment strategy: results of the NeuroThera Effectiveness and Safety Trial-1 (NEST-1) *Stroke.* **38** 1843–9

[29] Huisa B N *et al* 2013 Transcranial laser therapy for acute ischemic stroke: a pooled analysis of NEST-1 and NEST-2 *Int. J. Stroke.* **8** 315–20

[30] Zivin J A *et al* 2014 NeuroThera(R) Efficacy and Safety Trial-3 (NEST-3): a double-blind, randomized, sham-controlled, parallel group, multicenter, pivotal study to assess the safety and efficacy of transcranial laser therapy with the NeuroThera(R) Laser System for the treatment of acute ischemic stroke within 24 h of stroke onset *Int. J. Stroke.* **9** 950–5

[31] Zivin J A *et al* 2009 Effectiveness and safety of transcranial laser therapy for acute ischemic stroke *Stroke.* **40** 1359–64

[32] Oron A *et al* 2007 low-level laser therapy applied transcranially to mice following traumatic brain injury significantly reduces long-term neurological deficits *J. Neurotrauma.* **24** 651–6

[33] Wu Q *et al* 2012 Low-level laser therapy for closed-head traumatic brain injury in mice: effect of different wavelengths *Lasers Surg. Med.* **44** 218–26

[34] Ando T *et al* 2011 Comparison of therapeutic effects between pulsed and continuous wave 810-nm wavelength laser irradiation for traumatic brain injury in mice *PLoS One.* **6** e 26212

[35] Karu T I, Pyatibrat L V and Afanasyeva N I 2005 Cellular effects of low power laser therapy can be mediated by nitric oxide *Lasers Surg. Med.* **36** 307–14

[36] Wu Q *et al* 2012 Low-level laser therapy for closed-head traumatic brain injury in mice: effect of different wavelengths *Lasers Surg. Med.* **44** 218–226

[37] Ando T *et al* 2011 Comparison of therapeutic effects between pulsed and continuous wave 810-nm wavelength laser irradiation for traumatic brain injury in mice *PLoS ONE.* **6** e26212–26220

[38] Xuan W *et al* 2013 Transcranial low-level laser therapy improves neurological performance in traumatic brain injury in mice: effect of treatment repetition regimen *PLoS ONE.* **8** e 53454

[39] Quirk B J *et al* 2012 Near-infrared photobiomodulation in an animal model of traumatic brain injury: improvements at the behavioral and biochemical levels *Photomed. Laser Surg.* **30** 523–9

[40] Zhang Q *et al* 2014 Low-level laser therapy effectively prevents secondary brain injury induced by immediate early responsive gene X-1 deficiency *J. Cereb. Blood Flow Metab.*

[41] Dong T *et al* 2015 Low-level light in combination with metabolic modulators for effective therapy of injured brain *J. Cereb. Blood Flow Metab.*

[42] Naeser M A and Hamblin M R 2015 Traumatic brain injury: a major medical problem that could be treated using transcranial, red/near-infrared LED photobiomodulation *Photomed. Laser Surg.*

[43] McClure J 2011 The role of causal attributions in public misconceptions about brain injury *Rehabil. Psychol.* **56** 85–93

[44] Naeser M A *et al* 2011 Improved cognitive function after transcranial, light-emitting diode treatments in chronic, traumatic brain injury: two case reports *Photomed. Laser Surg.* **29** 351–8

[45] Naeser M A *et al* 2010 Improved language in a chronic nonfluent aphasia patient after treatment with CPAP and TMS *Cogn. Behav. Neurol.* **23** 29–38

[46] Naeser M A *et al* 2014 Significant improvements in cognitive performance post-transcranial, red/near-infrared light-emitting diode treatments in chronic, mild traumatic brain injury: open-protocol study *J. Neurotrauma.* **31** 1008–17

[47] Henderson T A and Morries L D 2015 SPECT perfusion imaging demonstrates improvement of traumatic brain injury with transcranial near-infrared laser phototherapy *Adv. Mind Body Med.* **29** 27–33

[48] De Taboada L *et al* 2011 Transcranial laser therapy attenuates amyloid-beta peptide neuropathology in amyloid-beta protein precursor transgenic mice *J. Alzheimers. Dis.* **23** 521-35

[49] Saltmarche A E, Ho K F, Hamblin M R and Lim L 2007 Significant improvement in cognition in mild to moderately severe dementia cases treated with transcranial plus intranasal photobiomodulation: Case Series Report *Photomed Laser Surg.* **35** 432–41

[50] Johnstone D M *et al* 2015 Turning on lights to stop neurodegeneration: the potential of near infrared light therapy in Alzheimer's and Parkinson's disease *Front. Neurosci.* **9** 500

[51] Mitrofanis J 2017 Why and how does light therapy offer neuroprotection in Parkinson's disease? *Neural Regen. Res.* **12** 574–575

[52] Bove J and Perier C 2012 Neurotoxin-based models of Parkinson's disease *Neuroscience.* **211** 51–76

[53] Shaw V E *et al* 2010 Neuroprotection of midbrain dopaminergic cells in MPTP-treated mice after near-infrared light treatment *J. Comp. Neurol.* **518** 25–40

[54] Peoples C *et al* 2012 Photobiomodulation enhances nigral dopaminergic cell survival in a chronic MPTP mouse model of Parkinson's disease *Parkinsonism Relat. Disord.* **18** 469–76

[55] Purushothuman S *et al* 2013 The impact of near-infrared light on dopaminergic cell survival in a transgenic mouse model of Parkinsonism *Brain Res.* **1535** 61–70

[56] Moro C *et al* 2014 Photobiomodulation inside the brain: a novel method of applying near-infrared light intracranially and its impact on dopaminergic cell survival in MPTP-treated mice *J. Neurosurg.* **120** 670–83

[57] Johnstone D M *et al* 2014 Indirect application of near infrared light induces neuroprotection in a mouse model of parkinsonism—an abscopal neuroprotective effect *Neuroscience.* **274** 93–101

[58] Reinhart F *et al* 2016 Intracranial application of near-infrared light in a hemi-Parkinsonian rat model: the impact on behavior and cell survival *J. Neurosurg.* **124** 1829–41

[59] El Massri N *et al* 2016 Near-infrared light treatment reduces astrogliosis in MPTP-treated monkeys *Exp. Brain Res.*

[60] Moro C *et al* 2016 Effects of a higher dose of near-infrared light on clinical signs and neuroprotection in a monkey model of Parkinson's disease *Brain Res.*

[61] Willner P 1997 Validity, reliability and utility of the chronic mild stress model of depression: a 10-year review and evaluation *Psychopharmacology.* **134** 319–29

[62] Anisman H and Matheson K 2005 Stress, depression, and anhedonia: caveats concerning animal models *Neurosci. Biobehav. Rev.* **29**(4-5) 525–46

[63] Wu X *et al* 2012 Pulsed light irradiation improves behavioral outcome in a rat model of chronic mild stress *Lasers Surg. Med.* **44** 227–32

[64] Salehpour F *et al* 2016 Therapeutic effects of 10-Hz pulsed wave lasers in rat depression model: a comparison between near-infrared and red wavelengths *Lasers Surg. Med.*

[65] Schiffer F *et al* 2009 Psychological benefits 2 and 4 weeks after a single treatment with near infrared light to the forehead: a pilot study of 10 patients with major depression and anxiety *Behav. Brain Funct.* **5** 46

[66] Cassano P *et al* 2015 Near-infrared transcranial radiation for major depressive disorder: proof of concept study *Psychiatry J.* **2015** 352979

Photomedicine and Stem Cells
The Janus face of photodynamic therapy (PDT) to kill cancer stem cells, and photobiomodulation
(PBM) to stimulate normal stem cells
Heidi Abrahamse and Michael R Hamblin

Chapter 9

PBM and stem cells in hair regrowth

9.1 Introduction to hair growth

A strand of hair principally consists of the cortex and the cuticle. The highly structured cortex provides mechanical strength and contains melanin granules providing pigmentation. The cuticle is the outer covering of protein with a single molecular layer of lipid that makes the hair repel water [1].

The hair follicle (HF) is a complex and dynamic mini-organ [2] embedded within the skin and is composed of different anatomical regions called the papilla, matrix, root sheath, and bulge [3]. There are between 250 000–500 000 HFs on the human scalp and as many as 5 000 000 on the whole body.

Hair growth is a cyclic process during which the HF moves sequentially from one phase to another [4] (figure 9.1). Anagen is the growth phase of the hair; catagen is the involuting or regressing phase; and telogen the resting or quiescent phase. Normally, up to 90% of the HF are in anagen phase, while 10%–14% are in telogen and only 1%–2% in catagen [5].

The signaling involved in the well-orchestrated process of hair growth and HF cycling is complex and incompletely understood [6]. The basic driving force is interaction between the mesenchymal and epithelial cell populations within the HF [7]. The most important mesenchymal cells in the HF reside within the dermal papilla and produce signals that control sequential cycling of the follicular epithelium [8]. This signaling leads to production of progenitor cells from the stem cells in the bulge area, and then these progenitor cells become transiently amplifying cells that expand downwards into the deep dermis, followed by differentiation into matrix cells that have the ability to produce the hair shaft, and its sheath. However in both humans and, in particular, other animals, the male and female sexes have very different hair phenotypes, which are governed by the influence of sex hormones [9]. Several growth factor families are involved in HF cycling [8], namely fibroblast growth factor (FGF), epidermal growth factor (EGF),

doi:10.1088/978-1-6817-4321-9ch9

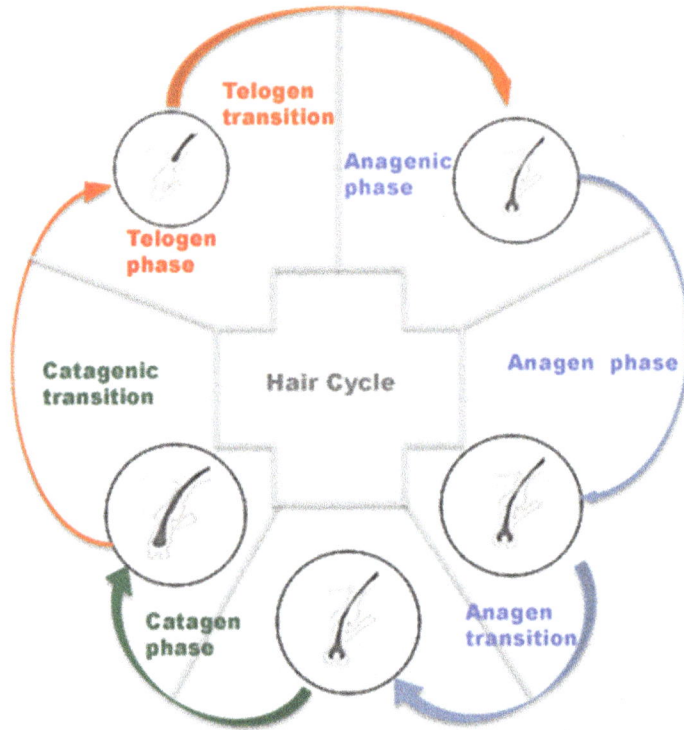

Figure 9.1. Hair cycle and its transitions. There are three main phases of the hair growth cycle; anagen, catagen, and telogen, with anagen further subdivided into proanagen, mesanagen, and metanagen.

hepatocyte growth factor (HGF), insulin-like growth factor (IGF)-I, and transforming growth factor-β (TGF-β) families, among others. Signal transducer and activator of transcription 3 (Stat3) is a latent cytoplasmic protein that conveys signals to the nucleus upon stimulation with cytokines/growth factors, leading to transcriptional activation of downstream genes that have the stat3 response element in their promoter region. Stat3 plays a critical role in HF cycling [10]. There is another, stat3-independent, pathway involving protein kinase C (PKC), but both pathways eventually lead to activation of phosphatidylinositol-4,5-bisphosphate 3-kinase (PI3K). Figure 9.2 shows some of the interactions and signaling that take place between the various cell types in the skin.

An interesting study by Hamanaka *et al* [11] showed that mitochondrial-generated reactive oxygen species (ROS) were critical mediators of cellular differentiation and HF morphogenesis by activating notch and β-catenin signals essential for epidermal differentiation and HF development [12].

9.2 Skin, hair follicles, and stem cells

Skin and its appendages carry out a number of critical functions necessary for animal survival. Skin protects animals from water loss, changes in temperature, solar radiation, external trauma, and infections by pathogens. Moreover, skin allows

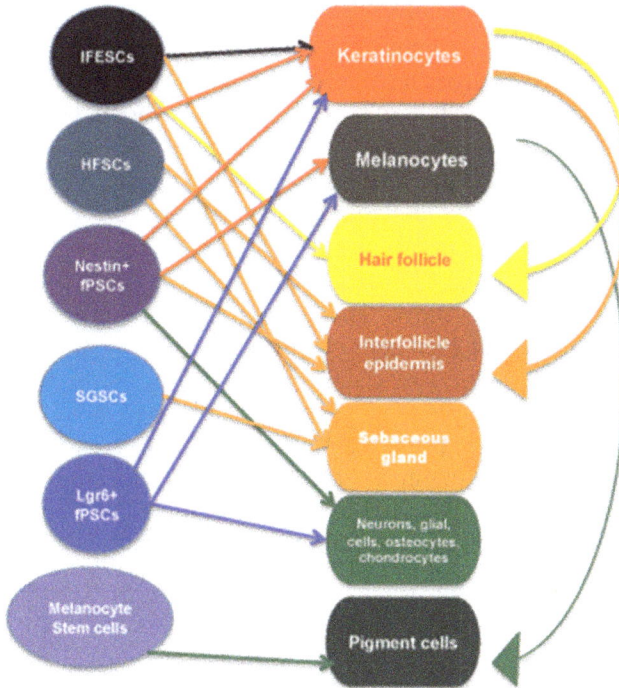

Figure 9.2. Interactions between stem cells, progenitor cells, and cells in and related to the skin. IFESCs, interfollicle epidermal stem cells; HFSCs, hair follicle stem cells; SGSCs, sebaceous gland stem cells; fPSCs, follicle nestin+ pluripotent stem cells. Lgr6+ fPSCs could be identical to fPSCs.

animals to perceive and interact with their environment through tactile senses such as touch, temperature, pressure, etc. In animals using environmental camouflage, the skin provides protection against predators, and it also serves a visual function for social and reproductive behavior. Therefore skin is one of the most critical organs and must be able to renew itself through the whole life-span.

Adult skin is composed of a diverse organized array of cells from different embryonic origins. In mammals, shortly after gastrulation, the neurectoderm cells at the embryonic surface become the epidermis, beginning as a single layer of unspecified progenitor cells. During development, this layer of cells differentiates to form the stratified epidermis (sometimes called interfollicular epidermis), the HFs, sebaceous glands, and apocrine (sweat) glands. Mesoderm-derived cells produce the collagen-secreting fibroblasts in the underlying dermis, the vasculature that supplies nutrients to the skin, *arrector pili* muscles that attach to each HF, the subcutaneous fat cells, and the variety of immune cells that infiltrate and reside in the skin. Neural crest-derived cells contribute to melanocytes, sensory nerve endings of the skin, and the dermis of the head. Overall, approximately 20 different cell types reside within the skin.

In the adult, many different types of stem cells function to replenish these various cell types in skin as it undergoes normal homeostasis or wound repair. Some stem cells (e.g. those that replenish lymphocytes) reside elsewhere in the body. Others (e.g. melanoblasts and epidermal stem cells) reside within the skin itself.

Cells with stem cell properties have recently been described in many integumental appendages, but the HF stands out as one of the best model systems for studying adult stem cells [3]. HF are accessible, well defined in terms of their developmental biology, and their stem cell populations are located in discrete compartments or niches. Since the landmark paper almost 25 years ago by Cotsarelis *et al* [13] highlighted the HF bulge as an important anatomical niche for HF epithelial stem cells in mice, much work directed at refining the location of stem cell compartments has been ongoing. Figure 9.3 shows the organization of the HF. Figure 9.4 shows a fluorescent labeling of adult stem cells on HFs.

9.3 Alopecia

Hair loss can be caused by any number of conditions, and can be classified as one out of a set of specific diagnoses, according to the American Society of Hair Loss Association. Some diagnoses have alopecia in their title, such as androgenic alopecia

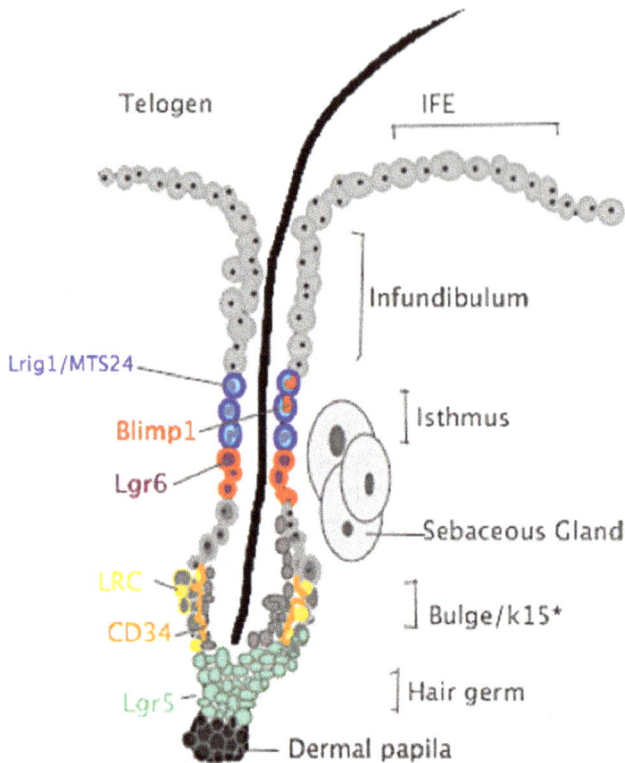

Figure 9.3. Schematic organization of the telogen-phase adult HF showing location of the stem cells. The stem cell populations are represented by their well-marked gene/protein expression or promoter activity: Lgr5 (hair germ and bulge), CD34 (bulge), LRC (bulge), Lgr6 (lower isthmus), Lrig1/MTS24 (isthmus), Blimp1 (sebaceous gland), and K15* (K15 promoter, bulge area).

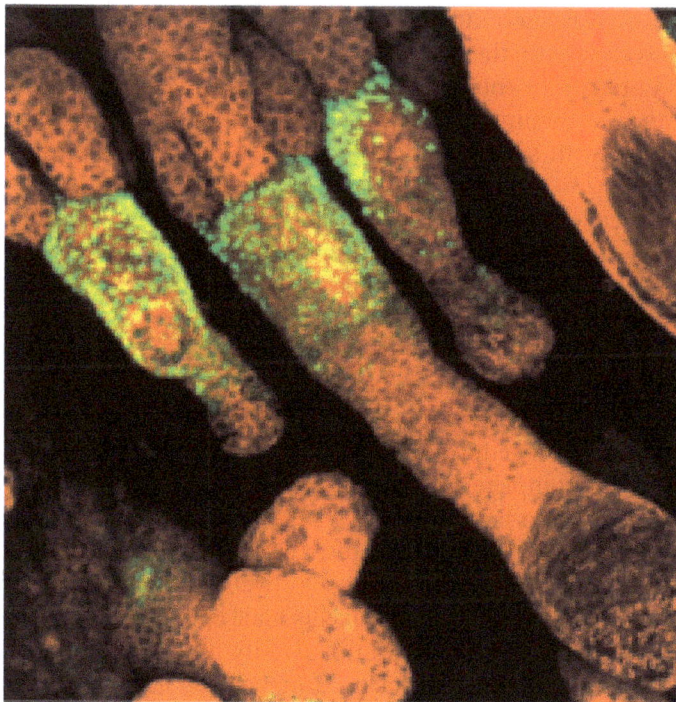

Figure 9.4. Fluorescent labeling of stem cells in a HF. In adult skin, each HF possesses a reservoir of stem cells. A green fluorescent protein labeled histone is first expressed in all the epithelial cells of the skin, but then its expression is turned off, the more rapidly dividing progeny dilute out the label, and only the stem cells retain it. The fluorescent red labels the membrane boundaries of the HF cells [14].

(AGA), alopecia areata (AA) or chemotherapy-induced alopecia (CIA), while others do not. AGA is by far the most common type of alopecia, characterized by progressive hair loss caused by androgenic miniaturization of the HF, and has a gradual increase in incidence by age. Although AGA is often called 'male pattern baldness' it can affect up to 70% of men and also 40% of women at some point in their lifetime.

Probably the most common non-AGA alopecias a dermatologist will see are telogen effluvium (TE), AA, and hair loss due to cosmetic over-processing. Other, more rare forms of hair loss may be difficult to diagnose, and some patients may wait months, even years for a correct diagnosis and undergo consultation with numerous dermatologists until they find one with knowledge of their condition. In addition, with rare diseases there is little motivation for research to be conducted and for treatments to be developed. Often, even when a correct diagnosis is made, a dermatologist can offer no known treatment for the condition.

TE is a particularly alarming form of alopecia [15]. The typical patient is a woman claiming to have always had a 'full head of hair' and reporting her hair to come out suddenly 'by the handful'. Patients are usually in good health, with no anorexia or nutrient deficiencies, possibly anxious, and have occasionally felt a painful or burning sensation at the scalp (trichodynia). Usually, the course of the disorder is chronic but intermittent, with apparent remissions followed by relapses. TE can be classified into three main categories: premature teloptosis, collective teloptosis, and premature entry into telogen (drug-induced TE, TE due to dietary deficiencies, and 'autoimmune' TE). The majority of TE patients are of the autoimmune type, with some analogies with AA, including the triggering role of emotional stress.

AGA is often called 'male pattern baldness', but the condition can affect large numbers of both men and women. Men typically present with hairline recession at the temples and balding at the vertex, while women normally suffer from diffuse hair thinning over the top of their scalp. Both genetic and environmental factors play a role, and the etiology remains incompletely understood.

Classic androgenic hair loss in males begins above the temples and vertex, or crown, of the scalp. As it progresses, only a thin rim of hair at the sides and rear of the head may remain (sometimes referred to as a 'Hippocratic wreath'), but AGA rarely progresses to complete baldness. The Hamilton–Norwood scale has been developed to grade AA in males [16].

Female AA has been colloquially referred to as 'female pattern baldness', although this pattern can occur in males as well. It more often causes diffuse thinning without hairline recession, and like its male counterpart rarely leads to total hair loss [17]. The Ludwig scale grades the severity of AA in females [18].

It is accepted that AA is caused by changes in the male steroid hormones known as androgens [19], and variants in the gene for the androgen receptor have been implicated in AGA [20]. Androgens are important in male sexual development around birth and at puberty. They regulate sebaceous glands, apocrine hair growth, and libido. With increasing age, androgens stimulate hair growth on the face, but suppress it at the temples and scalp vertex, a condition that has been referred to as the 'androgen paradox'. Men with AA typically have lower total testosterone, higher unbound/free testosterone, and higher free androgens, including dihydrotestosterone (DHT) [21]. The enzyme that transforms testosterone into DHT is known as 5-α-reductase (or 3-oxo-5α-steroid 4-dehydrogenase). There are three isoenzymes of 5-α-reductase: known as 5α-R1, 5α-R2, and 5α-R3. 5α-R1 is expressed about 50 times more in the scalp of adult males than in the scalp of a male fetus. Males with AGA have more 5α-R1 in the scalp HFs than males without AGA [21]. Therefore the prevailing hypothesis is that DHT is responsible for the miniaturization of HFs [22].

Adrenal steroids such as dehydroepiandrosterone can be converted to 5α-DHT by isolated HFs, which may provide an additional source of intrafollicular DHT. Elevated urinary dehydroepiandrosterone and serum dehydroepiandrosterone sulfate have been reported to be present in balding young men. These reports suggest that dehydroepiandrosterone sulfate may act as an important endocrine factor in the development of AGA [23].

The difference between AGA and CIA can be summarized as follows. Chemotherapy targets transiently amplifying progenitor cells but spares the mostly static stem cells. Therefore, the hairs regrow after the chemotherapy treatment is finished. In contrast, androgens inhibit Wnt signaling [24], which is required for the ability of the dermal papilla to induce hair cycling and regeneration [25]. This mechanism implies that targeting androgens as a therapy for AGA gives only a transient effect, while CIA can completely recover and hair growth becomes normal again.

The growth and dormancy of HFs have been related to the activity of IGF at the dermal papillae, which is affected by DHT [19]. Studies looking at serum levels of IGF-1 have shown it to be increased with vertex balding [26]. Earlier work looking at *in vitro* administration of IGF had no effect on HFs when insulin was present, but when insulin was absent IGF-1 caused follicle growth [27]. The effects on hair of IGF-I were found to be greater than IGF-II [28]. IGF-1 signaling controls the hair growth cycle and differentiation of hair shafts, possibly having an anti-apoptotic effect during the catagen phase. Mutations of the gene encoding IGF-1 result in Laron syndrome shortened and morphologically bizarre hair growth and alopecia [29]. IGF-1 is modulated by a IGF binding protein, which is produced in the dermal papilla [30].

DHT inhibits hair growth in mice by inhibiting IGF-1 at the dermal papilla [31]. The involvement of IGF signaling in the ongoing functioning of HF stem cells adds another angle to the effects of DHT on hair growth [32].

Extracellular histones inhibit hair shaft elongation and promote regression of HFs by decreasing IGF and alkaline phosphatase in transgenic mice [33]. Silencing P-cadherin, a HF protein at adherens junctions, decreases IGF-1 and increases TGF beta 2, although neutralizing TGF decreased catagenesis caused by loss of cadherin, suggesting additional molecular targets for therapy. P-cadherin mutants have short, sparse hair [34]. In addition, androgens enhance inducible nitric oxide synthase from occipital dermal papilla cells and stem cell factor for positive regulation of hair growth in beard and negative regulation of balding dermal papilla cells [35]. Other research suggests that the enzyme prostaglandin D2 synthase and its product, prostaglandin D2 (PGD2) in HFs, could contribute to hair loss in AGA [36].

Recently, a newly discovered class of tiny non-coding RNAs named microRNAs have been discovered to play a role in regulating gene expression, and have led to the new field of 'regulomics' [37]. These tiny bits of transcriptome fine-tune the expression of nearly one-third of the genes in mammals [38]. MicroRNAs affect skin development [39], and play a role in HF morphogenesis, maintenance, and cycling (controlling proliferation, growth arrest, and apoptosis) [40]. MicroRNAs have also been shown to play a role in the pathogenesis of AGA, with higher expression in the dermal papillae of HFs from AGA patients versus normal HFs [41].

Reduced blood flow and oxygen pressure in the balding areas are also involved in male pattern baldness. It has been shown that decreased blood flow and lower partial pressure of oxygen occurs in the balding scalp compared with non-balding areas and with controls [42]. The decreases found in partial pressure of oxygen and blood flow are approximately 40% and 62%, respectively. However, it is unlikely that these decreases in partial pressure of oxygen or blood flow are the root cause of hair loss. Only a fraction of the blood flow to the skin is used for metabolic needs and the rest is for temperature regulation. Furthermore, even for metabolically very active cells, the critical partial pressure of oxygen for impairment of mitochondrial function is less than 5 mmHg and about 0.1 mmHg for hypoxic cell death [43].

9.4 PBM for alopecia

In order to test the effect of linear polarized infrared irradiation in the treatment of AA, a study was conducted with 15 patients (6 men, 9 women) using Super Lizer™, a medical instrument emitting polarized pulsed linear light with a high output (1.8 W) of infrared radiation (600–1600 nm) that is capable of penetrating into deep subcutaneous tissue [44]. The scalp was irradiated for 3 min either once every week or once every other week until vellus hair regrowth in at least 50% of the affected area was observed. Additionally carpronium chloride 5% was applied topically twice daily to all the lesions in combination with oral antihistamines, cepharanthin, and glycyrrhizin (extracts of Chinese medicine herbs) [44]. As a result of this study, in 47% of the patients' hair growth occurred 1.6 months earlier in irradiated areas than in non-irradiated areas [44]. However, one year after irradiation, all the lesions disappeared; hair density, length and diameter of hair shafts were the same both in irradiated and non-irradiated lesions; suggesting that low-level light therapy (LLLT) only accelerates the process of hair regrowth in AA patients. It is worth mentioning that the method for assessment of hair regrowth, density, and thickness was not clearly stated, which was one of the main limitations of this study.

Using 655 nm red light and 780 nm infrared light once a day for 10 min, 24 male AGA patients were treated and evaluated by a group of investigators [45]. Evaluation was performed via global photography and phototrichogram [45]. Following 14 weeks of treatment, the density of hairs and anogen/telogen ratio on both the vertex and occiput was significantly increased, and 83% of the patients reported to be satisfied with the treatment [45].

Satino *et al* tested the efficacy of LLLT on hair growth and tensile strength on 28 male and 7 female AGA patients [46]. Each patient was given a HairMax LaserComb® 655 nm, to use at home for 6 months for 5–10 min every other day [46]. Tensile strength was measured by VIP HairOSCope (Belson Imports, Hialeah, FL) through removal of three typical terminal hairs from a one square centimeter area. Hair count was performed within a 1 cm square space created within a mold that was prepared around the area of greatest alopecia. A surgical hook and magnification were used while counting the number of hair. In terms of hair tensile strength, the results revealed greater improvement in the vertex area for males and temporal area for females, however both sexes benefited in all areas significantly [46]. In terms of hair count, both sexes and all areas had substantial improvement (for the temporal area: 55% in women, 74% in men; in the vertex area: 65% in women, 120% in men) with the vertex area in males having the best outcome [46]. The LaserComb device was tested by Leavitt *et al* in a double-blind, sham device-controlled, multicenter, 26 week trial, randomized study among 110 male AGA patients [47]. Patients used the device three times per week for 15 min for a total of 26 weeks [47]. A significantly greater increase in mean terminal hair density compared to subjects in the sham device group was reported [47]. Significant improvements in overall hair regrowth, slowing of hair loss, thicker feeling hair, better scalp health, and hair shine were also demonstrated in terms of patients' subjective assessment at 26 weeks over baseline [47].

Recently, a double-blind randomized controlled trial by Lanzafame *et al* using a helmet containing 21 5 mW lasers and 30 LEDs (655 ± 5 nm, 67.3 J cm^{-2} 25 min^{-1} treatment) every other day for 16 weeks reported 35% increase in hair growth among male AGA patients [48]. Another recent study by Kim *et al* designed a 24 week randomized, double-blind, sham device-controlled, multicenter trial among both male and female AGA patients in order to investigate the efficacy of a helmet type LLLT device combining a 650 nm laser with 630 and 660 nm LEDs (total energy density 92.15 mW cm^{-2}, 47.90 J cm^{-2} for 18 min). Even though mean hair thickness (12.6 ± 9.4 versus 3.9 ± 7.3 in the control group, $p = 0.1$) and hair density (17.2 ± 12.1 versus –2.1 ± 18.3 in the control group, $p = 0.003$) increased significantly in the treatment group, there was no prominent difference in global appearance between the two groups [49]. Findings from a different study by Avram and Rogers were in accordance with these results, where LLLT increased hair count and shaft diameter, however blinded global images did not support these observations [50].

LLLT has demonstrated a remarkably low incidence of adverse effects when it has been used over 50 years for diverse medical conditions and in a variety of anatomical sites. In the specific area of LLLT for hair growth, the only adverse report in humans was the temporary onset of TE developing in the first 1–2 months after commencing LaserComb treatment [46], but disappearing on continued application. Some other possible considerations are the presence of dysplastic or malignant lesions on the scalp which could be stimulated to grow by the proliferative effects of LLLT [51].

References

[1] Draelos Z K 1991 Hair cosmetics *Dermatol. Clin.* **9** 19–27

[2] Schneider M R, Schmidt-Ullrich R and Paus R 2009 The hair follicle as a dynamic miniorgan *Curr. Biol.* **19** R132–42

[3] Mokos Z B and Mosler E L 2014 Advances in a rapidly emerging field of hair follicle stem cell research *Coll. Antropol.* **38** 373–8

[4] Paus R 2006 Therapeutic strategies for treating hair loss *Drug Discov. Today: Ther. Strateg.* **3** 101–110

[5] Sennett R and Rendl M 2012 Mesenchymal-epithelial interactions during hair follicle morphogenesis and cycling *Semin. Cell. Dev. Biol.* **23** 917–27

[6] Oro A E and Scott M P 1998 Splitting hairs: dissecting roles of signaling systems in epidermal development *Cell.* **95** 575–8

[7] Jahoda C A and Reynolds A J 1996 Dermal-epidermal interactions. Adult follicle-derived cell populations and hair growth *Dermatol. Clin.* **14** 573–83

[8] Peus D and Pittelkow M R 1996 Growth factors in hair organ development and the hair growth cycle *Dermatol. Clin.* **14** 559–72

[9] Mayer J A, Chuong C M and Widelitz R 2004 Rooster feathering, androgenic alopecia, and hormone-dependent tumor growth: what is in common? *Differentiation.* **72** 474–88

[10] Sano S *et al* 2000 Two distinct signaling pathways in hair cycle induction: Stat3-dependent and -independent pathways *Proc. Natl Acad. Sci. USA.* **97** 13824–9

[11] Hamanaka R B *et al* 2013 Mitochondrial reactive oxygen species promote epidermal differentiation and hair follicle development *Sci. Signal.* **6** ra8

[12] Aubin-Houzelstein G 2012 Notch signaling and the developing hair follicle *Adv. Exp. Med. Biol.* **727** 142–60

[13] Cotsarelis G, Sun T T and Lavker R M 1990 Label-retaining cells reside in the bulge area of pilosebaceous unit: implications for follicular stem cells, hair cycle, and skin carcinogenesis *Cell.* **61** 1329–37

[14] Fuchs E, Tumbar T and Guasch G 2004 Socializing with the neighbors: stem cells and their niche *Cell.* **116** 769–78

[15] Rebora A 2014 Telogen effluvium revisited *G. Ital. Dermatol. Venereol.* **149** 47–54

[16] Guarrera M *et al* 2009 Reliability of hamilton-norwood classification *Int. J. Trichol.* **1** 120–2

[17] Herskovitz I and Tosti A 2013 Female pattern hair loss *Int. J. Endocrinol. Metab.* **11** e9860

[18] Guarrera M, Semino M T and Rebora A 1997 Quantitating hair loss in women: a critical approach *Dermatology.* **194** 12–6

[19] Kaufman K D 2002 Androgens and alopecia *Mol. Cell. Endocrinol.* **198** 89–95

[20] Levy-Nissenbaum E *et al* 2005 Confirmation of the association between male pattern baldness and the androgen receptor gene *Eur. J. Dermatol.* **15** 339–40

[21] Crabtree J S *et al* 2010 A mouse model of androgenetic alopecia *Endocrinology.* **151** 2373–80

[22] Kaliyadan F, Nambiar A and Vijayaraghavan S 2013 Androgenetic alopecia: an update *Ind. J. Dermatol. Venereol. Leprol.* **79** 613–25

[23] Poor V, Juricskay S and Telegdy E 2002 Urinary steroids in men with male-pattern alopecia *J. Biochem. Biophys. Methods.* **53** 123–30

[24] Leiros G J, Attorresi A I and Balana M E 2012 Hair follicle stem cell differentiation is inhibited through cross-talk between Wnt/beta-catenin and androgen signalling in dermal papilla cells from patients with androgenetic alopecia *Br. J. Dermatol.* **166** 1035–42

[25] Dong L *et al* 2014 Treatment of MSCs with Wnt1a-conditioned medium activates DP cells and promotes hair follicle regrowth *Sci. Rep.* **4** 5432

[26] Platz E A *et al* 2000 Vertex balding, plasma insulin-like growth factor 1, and insulin-like growth factor binding protein 3 *J. Am. Acad. Dermatol.* **42** 1003–7

[27] Kamiya T *et al* 1998 Hair follicle elongation in organ culture of skin from newborn and adult mice *J. Dermatol. Sci.* **17** 54–60

[28] Panchaprateep R and Asawanonda P 2014 Insulin-like growth factor-1: roles in androgenetic alopecia *Exp. Dermatol.* **23** 216-8

[29] Lurie R, Ben-Amitai D and Laron Z 2004 Laron syndrome (primary growth hormone insensitivity): a unique model to explore the effect of insulin-like growth factor 1 deficiency on human hair *Dermatology.* **208** 314–8

[30] Batch J A, Mercuri F A and Werther G A 1996 Identification and localization of insulin-like growth factor-binding protein (IGFBP) messenger RNAs in human hair follicle dermal papilla *J. Invest. Dermatol.* **106** 471–5

[31] Zhao J, Harada N and Okajima K 2011 Dihydrotestosterone inhibits hair growth in mice by inhibiting insulin-like growth factor-I production in dermal papillae *Growth Horm. IGF Res.* **21** 260–7

[32] Li J *et al* 2014 Exogenous IGF-1 promotes hair growth by stimulating cell proliferation and down regulating TGF-beta1 in C57BL/6 mice *in vivo Growth Horm IGF Res.* **24** 89–94

[33] Shin S H *et al* 2012 Extracellular histones inhibit hair shaft elongation in cultured human hair follicles and promote regression of hair follicles in mice *Exp. Dermatol.* **21** 956–8

[34] Samuelov L *et al* 2012 P-cadherin regulates human hair growth and cycling via canonical Wnt signaling and transforming growth factor-beta2 *J. Invest. Dermatol.* **132** 2332–41

[35] Inui S and Itami S 2013 Androgen actions on the human hair follicle: perspectives *Exp. Dermatol.* **22** 168–71

[36] Garza L A *et al* 2012 Prostaglandin D2 inhibits hair growth and is elevated in bald scalp of men with androgenetic alopecia *Sci. Transl. Med.* **4** 126ra34

[37] Ambros V 2004 The functions of animal microRNAs *Nature.* **431** 350–5

[38] Lewis B P, Burge C B and Bartel D P 2005 Conserved seed pairing, often flanked by adenosines, indicates that thousands of human genes are microRNA targets *Cell.* **120** 15–20

[39] Sand M *et al* 2009 MicroRNAs and the skin: tiny players in the body's largest organ *J. Dermatol. Sci.* **53** 169–75

[40] Andl T *et al* 2006 The miRNA-processing enzyme dicer is essential for the morphogenesis and maintenance of hair follicles *Curr. Biol.* **16** 1041–9

[41] Goodarzi H R *et al* 2010 MicroRNAs take part in pathophysiology and pathogenesis of male pattern baldness *Mol. Biol. Rep.* **37** 2959–65

[42] Goldman B E, Fisher D M and Ringler S L 1996 Transcutaneous PO2 of the scalp in male pattern baldness: a new piece to the puzzle *Plast. Reconstr. Surg.* **97** 1109-16 1117

[43] Klemp P, Peters K and Hansted B 1989 Subcutaneous blood flow in early male pattern baldness *J. Invest. Dermatol.* **92** 725–6

[44] Yamazaki M *et al* 2003 Linear polarized infrared irradiation using Super Lizer is an effective treatment for multiple-type alopecia areata *Int. J. Dermatol.* **42** 738–40

[45] Kim S S, Park M W and Lee C J 2007 Phototherapy of androgenetic alopecia with low level narrow band 655-nm red light and 780-nm infrared light *J. Am. Acad. Dermatol.* AB112

[46] Satino J L and Markou M 2003 Hair regrowth and increased hair tensile strength using the HairMax LaserComb for low-level laser therapy *Int. J. Cosmet. Surg. Aesth. Dermatol.* **5** 113–117

[47] Leavitt M *et al* 2009 HairMax LaserComb laser phototherapy device in the treatment of male androgenetic alopecia: a randomized, double-blind, sham device-controlled, multi-centre trial *Clin. Drug Investig.* **29** 283–92

[48] Lanzafame R J, Blanche R R, Bodian A B, Chiacchierini R P, Fernandez-Obregon A and Kazmirek E R 2013 The growth of human scalp hair mediated by visible red light laser and LED sources in males *Lasers Surg Med.* **45** 487–95

[49] Kim H *et al* 2013 Low-level light therapy for androgenetic alopecia: a 24-week, randomized, double-blind, sham device-controlled multicenter trial *Dermatol. Surg.*

[50] Avram M R and Rogers N E 2009 The use of low-level light for hair growth: part I *J. Cosmet. Laser Ther.* **11** 110–7

[51] Frigo L *et al* 2009 The effect of low-level laser irradiation (In–Ga–Al–AsP—660 nm) on melanoma *in vitro.* and *in vivo BMC Cancer* **9** 404

Photomedicine and Stem Cells
The Janus face of photodynamic therapy (PDT) to kill cancer stem cells, and photobiomodulation
(PBM) to stimulate normal stem cells
Heidi Abrahamse and Michael R Hamblin

Chapter 10

Conclusion

The discovery of stem cells was one of the most important events in all of biomedicine. Stem cells are critically important for tissue regeneration, repair, and healing. Conversely cancer stem cells (CSCs) have a critically important role to play in many forms of cancer (malignant glioma, breast cancer, colon cancer, ovarian cancer, pancreatic cancer, leukemia, multiple myeloma, and skin cancer). The Janus nature of stem cells means that they can be highly beneficial in diseases where healing and regeneration are required, but conversely CSCs can be highly damaging in cancers, being responsible for tumor progression, recurrence, and development of resistance to chemotherapy. This dual nature of stem cells is counterbalanced by the dual nature of therapeutic light in photomedicine. As shown in previous chapters, when light is combined with properly chosen photosensitizers, CSCs can be efficiently killed by photodynamic therapy (PDT), especially when efforts are undertaken to inhibit multi-drug efflux pumps such as ABCG2.

On the other hand, photobiomodulation (PBM) can have remarkable benefits in stimulating the mobilization, proliferation, and differentiation of stem cells. The impressive results from Uri Oron (discussed in chapter 5) have shown that delivering near-infrared light to the bone marrow in the leg bones can have better therapeutic effects on distant organs such as the heart, kidney, and brain, compared to delivering light to the damaged organs themselves. A large number of investigators have attempted to use PBM to pre-condition various types of stem cells *in vitro* before implantation *in vivo* [1]. This has often taken the form of spheroids formed *in vitro* from adipose-derived stem cells (ADSCs) that can be implanted into experimental animals to stimulate angiogenesis in hind-limb ischemia [2] or skin flap ischemia [3]. Here the stem cells are proposed to differentiate into endothelial cells to allow growth of new blood vessels. This approach can also be used to enhance wound healing when the PBM-treated ADSC spheroids are implanted into the wound bed [4]. Another similar application is for the repair of bone defects [5].

The predominant role of stem cells in embryogenesis and development has not been much discussed in this book. The fact that embryos have the largest concentrations of stem cells compared to any other human tissue or organ could suggest that PBM may have important effects on human development. However, there is a completely understandable reluctance to suggest that PBM could be employed on developing embryos. This is of course because the entire medical profession is incredibly wary of repeating anything even remotely similar to the thalidomide disaster of the 1960s. This has been described as 'one of the darkest episodes in pharmaceutical research history'. Thalidomide was marketed as a mild sleeping pill safe even for pregnant women, however, it caused thousands of babies worldwide to be born with malformed limbs [6]. It was not until 50 years later that the mechanism of this horrific effect was understood [7]. Women usually took the drug at about five to nine weeks into their pregnancy to combat morning sickness, a specific window that is crucial to the formation of limbs of the developing fetus. The blood vessels involved in this process, are still at an immature stage when they rapidly change and expand to accommodate the outgrowing limb. The anti-angiogenic effect of thalidomide was later used as an ingredient of some cancer therapies (especially for multiple myeloma) [8, 9].

At present the application of PBM to the abdomen in pregnant women is an official contraindication by the World Association of Laser Therapy (WALT) [10]. In the future however, the effects of PBM on fetal development may be investigated in animal models. Of some to relevance to this idea is a paper from Japan that reports the use of PBM in female subjects with severe infertility [11]. Oshiro applied 830 nm laser to the neck (average 21 sessions) in 701 women who had failed to become pregnant for up to 9 years using standard assisted reproduction techniques. Pregnancy was achieved in 22% of the subjects, more than half of whom gave birth to healthy babies. While the involvement of stem cells in this process is unclear, it was clearly a systemic effect as PBM was applied to the neck.

As more knowledge is accumulated concerning the effects of PBM on stem cells, it is expected that the range of diseases and injuries that can be treated will expand considerably. PDT may also be applied for the destruction of CSCs, thus improving the overall results of cancer therapy.

References

[1] Liu Y and Zhang H 2016 Low-level laser irradiation precondition for cardiac regenerative therapy *Photomed. Laser Surg.* **34** 572–579

[2] Park I S, Chung P S and Ahn J C 2017 Adipose-derived stem cell spheroid treated with low-level light irradiation accelerates spontaneous angiogenesis in mouse model of hindlimb ischemia *Cytotherapy*

[3] Park I S, Chung P S and Ahn J C 2016 Angiogenic synergistic effect of adipose-derived stromal cell spheroids with low-level light therapy in a model of acute skin flap ischemia *Cells Tissues Organs* **202** 307–318

[4] Park I S, Chung P S and Ahn J C 2015 Adipose-derived stromal cell cluster with light therapy enhance angiogenesis and skin wound healing in mice *Biochem. Biophys. Res. Commun.* **462** 171–7

[5] Fekrazad R *et al* 2015 The effects of combined low level laser therapy and mesenchymal stem cells on bone regeneration in rabbit calvarial defects *J. Photochem. Photobiol.* **B 151** 180–5

[6] Vargesson N 2015 Thalidomide-induced teratogenesis: history and mechanisms *Birth Defects Res. C* **105** 140–56

[7] Ledford H 2009 How thalidomide makes its mark Nature News available from www.nature.com/news/2009/090511/full/news.2009.462.html

[8] Mercurio A *et al* 2017 A mini-review on thalidomide: chemistry, mechanisms of action, therapeutic potential and anti-angiogenic properties in multiple myeloma *Curr. Med. Chem.*

[9] Adlard J W 2000 Thalidomide in the treatment of cancer *Anticancer Drugs* **11** 787–91

[10] Navratil L and Kymplova J 2002 Contraindications in noninvasive laser therapy: truth and fiction *J. Clin. Laser Med. Surg.* **20** 341–3

[11] Ohshiro T 2012 Personal overview of the application of LLLT in severely infertile Japanese females *Laser Ther.* **21** 97–103